I0008910

Luminos is the Open Access monograph publishing program from UC Press. Luminos provides a framework for preserving and reinvigorating monograph publishing for the future and increases the reach and visibility of important scholarly work. Titles published in the UC Press Luminos model are published with the same high standards for selection, peer review, production, and marketing as those in our traditional program. www.luminosoa.org

This book is freely available in an open access edition thanks to TOME (Toward an Open Monograph Ecosystem)—a collaboration of the Association of American Universities, the Association of University Presses, and the Association of Research Libraries—and the generous support of the University of California—Davis. Learn more at the TOME website, available at: openmonographs.org.

Where Truth Lies

The publisher and the University of California Press Foundation gratefully acknowledge the generous support of the Ahmanson Foundation Endowment Fund in Humanities.

Where Truth Lies

Digital Culture and Documentary Media after 9/11

———

Kris Fallon

UNIVERSITY OF CALIFORNIA PRESS

University of California Press, one of the most distinguished university presses in the United States, enriches lives around the world by advancing scholarship in the humanities, social sciences, and natural sciences. Its activities are supported by the UC Press Foundation and by philanthropic contributions from individuals and institutions. For more information, visit www.ucpress.edu.

University of California Press
Oakland, California

Suggested citation: Fallon, K. *Where Truth Lies: Digital Culture and Documentary Media after 9/11*. Oakland: University of California Press, 2019. DOI: https://doi.org/10.1525/luminos.80

Library of Congress Cataloging-in-Publication Data

Names: Fallon, Kris, 1976– author.
Title: Where truth lies : digital culture and documentary media after 9/11 / Kris Fallon.
Description: Oakland, California : University of California Press, [2019] | Includes bibliographical references and index. |
Identifiers: LCCN 2019018762 (print) | ISBN 9780520300934 (pbk. : alk. paper) | ISBN 9780520972117 (ebook)
Subjects: LCSH: Digital media—Political aspects—United States—21st century. | Documentary mass media—United States—21st century. | Mass media—Objectivity—United States—21st century. | Online social networks—Political aspects—21st century.
Classification: LCC P95.82.U6 F35 2019 (print) | LCC P95.82.U6 (ebook) | DDC 302.23/10973—dc23
LC record available at https://lccn.loc.gov/2019018762
LC ebook record available at https://lccn.loc.gov/2019980041

28 27 26 25 24 23 22 21 20 19
10 9 8 7 6 5 4 3 2 1

For Alisa, Keaton, and Harper, whose love is the truest truth I know

CONTENTS

ILLUSTRATIONS

ACKNOWLEDGMENTS

Like many books, this one may have a single author's name on the cover but owes its existence to many others. While the final form of the book took shape over the last two years, the questions that it seeks to address have been with me in one form or another across many years, countless conversations, and four institutions.

My thinking on documentary aesthetics and the capacity of moving-image media and digital technology to explore the world has been indelibly shaped by many of the wonderful teachers and mentors with whom I have been fortunate enough to work over the years, including Marina Goldovskaya, Katherine Hayles, Jeffrey Decker, Randy Rutsky, Aaron Kerner, Jenny Lau, Kaja Silverman, Anton Kaes, Hubert Dreyfus, and David Bates. Particular thanks go to the dissertation committee on whom many of these ideas were first foisted: Kristen Whissel, Jeffrey Skoller, Ken Goldberg, Martin Jay, and the fearless chair of that committee, Linda Williams, who set a model for critical thinking and scholarly mentorship that I find myself striving to emulate every day with my own students. Like the field itself, my work on documentary in particular has been shaped immeasurably by Bill Nichols, whose books, seminars, office hours, and ultimate generosity as an adviser and scholar have been fundamental in my own intellectual path.

I am further grateful for the friendship and intellectual challenge posed by many of my colleagues in the wider field whose influence on this particular project has been both intangible and yet instrumental: Alenda Chang, Chris Goetz, Amanda Phillips, Brooke Belisle, Doug Cunningham, Jaimey Baron, Selmin Kara, Jen Schradie, Kris Paulsen, Tung-Hui Hu, Allison Fish, Alessandro Delfanti, Laura Horak, Erika Balsom, and Paige Sarlin. Damon Young's semioccasional refrain "What's happening with the book?" along with his example of tireless, good-natured productivity provided an important source of accountability. Ilona Hongisto's advice on framing the final chapter was invaluable, as was her friendship throughout the entire process. Particular thanks go to both Ben Stork and Kevin McDonald, who through countless conversations and tireless revisions provided detailed feedback on virtually every part of the manuscript. Good friends don't always make good readers, but in Ben and Kevin I was lucky enough to find both.

Various parts of the text found their first iteration on panels at conferences including SCMS, Visible Evidence, and others, and I am particularly grateful for the input of fellow copanelists, chairs, and attendees whose chance remarks, questions, and critiques on these occasions have strengthened the arguments made here; they include Lisa Parks, Richard Grusin, Jonathan Kahana, Finn Brunton, Kelly Gates, Julia Lesage, and Elizabeth Cowie. Special thanks go to Ted Nannicelli, Marguerite La Caze, and Tom O'Regan for the opportunity to present parts of this at the "(New) Visuality: Ethics and Aesthetics" symposium at the University of Queensland, and to the many generous interlocutors I encountered there, including Mette Hjort, Robert Sinnerbrink, and Damian Cox.

I would also like to thank my many colleagues at UC Davis both for their support and input on bringing the book project to completion and for further building UC Davis into an amazing place to think about politics and digital media in its many permutations. First and foremost are the two figures whom I consider to be the chief architects of this transformation—Colin Milburn and Kriss Ravetto—with whom I was fortunate enough to collaborate on the Mellon Digital Cultures initiative and who both provided fundamental input on this manuscript in various ways. Thanks also to Jaimey Fisher, Tim Lenoir, Stephanie Boluk, Patrick Lemieux, Joe Dumit, Lynette Hunter, Molly McCarthy, and Mario Biagioli for insights large and small, including simply according me the time and space to write and bring the project to fruition. Jonathan Doucette, Colin Johnson, and Emelie Mahdavian provided useful feedback on several chapters, and Andrea Miller provided invaluable summer research assistance on the entire manuscript. Thanks finally to Eric Smoodin for astute guidance on the publication process and for steering me toward UC Press.

Heartfelt thanks to my editor, Raina Polivka, who has been a stalwart force guiding this project from an unfinished manuscript through the twists and turns of the publication process with patience and confidence in the contribution that the book makes to the field. Dore Brown and Carl Walesa were also tremendously helpful patiently pulling together and copyediting the unusual array of digital and analog material that this manuscript relied. I am further grateful to Alexandra Juhasz and an anonymous reader who both offered generous and insightful input on the framing and scope of the arguments contained here. And particular thanks to Jen Malkowski for extensive and insightful notes on virtually every chapter; this project is unquestionably stronger because of her input. Thanks also to Mackenzie Smith and the University of California, Davis Library and the Dean of Letters & Science for the TOME grant that made the open-access publication of the book possible.

Lastly, I would like to express unending gratitude to the family and friends who have offered a deep well of support and patience as I took time on nights, weekends, and holidays to work on this over the last few years. Thanks to my dad, Michael, and my siblings Jason, AJ, and Ashley, and to my many aunts, uncles, and cousins, all of whom have supported and expressed continual faith in this project. Thanks also to my grandparents for an unending curiosity in what I'm "up to" now. Thanks also to my mom, without whom this and so many other things in life simply wouldn't have been possible. I would also like to thank Jonas Ball for convincing me to take a documentary film class many years ago at UCLA and Tom Flynn for a series of engaging debates on film, politics, and many other topics. Final and biggest thanks to my wife, Alisa, who patiently and enthusiastically read through countless drafts at various stages, and to my sons Keaton and Harper, who push me every day to see the world a little differently.

PREFACE

In 2016, when real-estate heir and reality-television figure Donald Trump was unexpectedly elected president of the United States, media were to blame. Mainstream media (the news) had missed a silent majority of working-class voters in their focus on the opinions and preferences of coastal elites. Marginal media (fake news) had manufactured lies and manipulated low-information voters. Quantitative media (data) had used faulty models and outdated polling techniques to predict the outcome. Social media (Twitter) had been weaponized to manipulate the news cycle and bully and intimidate critics. Deleted, hacked, and ultimately leaked media (e-mails) had cast doubt on an otherwise trustworthy candidate. Even the electoral college—a sort of political medium designed to transmit and translate the will of the voters into the constitutional form of elected office—had failed to accurately reflect and communicate the choice of a majority of voters.

While the full impact of Trump's election will take many years to play out, the initial surprise—some might say shock—it generated reveals a great deal about the relationship between politics and media (or politics as media) in contemporary American culture. The first revelation is the vast quantity and heterogeneity of information sources. Data, images, private messages, public proclamations, professional insiders, and renegade outsiders were all deemed credible in some context for some audience. Media have perhaps never before been so numerous or so diverse. The second point is that in spite of this variety, all of these forms are still considered nonfiction media. For the audiences they attract, they engender a degree of faith in their ability to accurately reflect reality. Simply put, they can tell the truth. They are, in other words, documentary media. And finally, of course, they were all wrong.

The text that follows will explain how the media landscape that produced this particular cultural and political event came to be over the prior fifteen to twenty years. My focus is not on Trump or the 2016 election specifically, but rather on the climate of political polarization that emerged in wake of the terrorist attacks of September 11, 2001; the culture of innovation that emerged with widespread access to digital technology in the late 1990s; and the persistence of a faith in nonfiction media to mobilize citizens that dates back to the earliest photographs and takes off in the early twentieth century. I will demonstrate that our current period of rapid evolution in documentary or nonfiction media has several historical precedents (chapter 1), and that its impact can be seen in a variety of contemporary forms including film (chapter 2), social media (chapter 3), video games (chapter 4), and data visualization (chapter 5). The final chapter will place these same forces—technology, politics, and nonfiction media—into the context of the 2016 election by looking at the role of conspiracy theories and midstream news outlets in the rise of Donald Trump. Moments of political rupture like the one we are living through produce a need for new forms of media capable of expressing and reflecting competing sets of values about what is or ought to be in our shared cultural life. This search to discover new forms of media capable of expressing these shared values—to determine, in other words, where truth lies—is what this book is about.

1

Seeing in the Dark

We also have to work, though, sort of the dark side, if you will. We've got to spend time in the shadows in the intelligence world. . . . It is a mean, nasty, dangerous dirty business out there, and we have to operate in that arena.

—DICK CHENEY, *MEET THE PRESS*, SEPTEMBER 16, 2001

Here in the pre-dawn darkness of Afghanistan, we can see the light of a new day on the horizon. . . . [T]hrough dark days we have drawn strength from . . . the ideals that have guided our nation and lit the world: a belief that all people are created equal, and deserve the freedom to determine their destiny. That is the light that guides us still.

—BARACK OBAMA, *PRESS CONFERENCE*, MAY 1, 2012

It's all working out. Just remember, everything that you're seeing, everything that you're reading, it's not what's happening.

—DONALD TRUMP, *SPEAKING TO VETERANS*, JULY 24, 2018

The statements in this chapter's epigraph punctuate a dramatic period in the history of the United States. In the early days of a postmillennium presidency, just two weeks after the terrorist attacks of September 11, 2001, Vice President Dick Cheney presaged that the country would be forced to "work . . . the dark side" and "spend time in the shadows" to wage a battle for which there would be "no end date." Signaling an end to the (relative) peace and prosperity of the prior decade, this new world was "mean, nasty, dirty dangerous," and survival in this landscape would require "any means at our disposal." Though he was referring to the government's planned approach to dealing with terrorist threats, in hindsight his remarks foretell the long period of deep political turmoil and conflict over events yet to come, events that included revelations of secret prisons, torture, human rights abuses, over a hundred thousand civilian causalities, two wars abroad, and an unprecedented erosion of civil liberties for average citizens at home. In the midst of this political conflict, the country was further shaken by the worst economic crisis since the Great Depression.

Just over a decade later, on the anniversary of the death of Osama bin Laden, however, President Barack Obama reassured the public that this period of darkness was over. The country had emerged from the "dark cloud of war" and was basking in "the light of a new day." Finally, we could rekindle the ideals that "have guided our nation and lit the world." The optimism of Obama's rhetoric notwithstanding, the end of the war that his comments foreshadowed has yet to arrive. Although it had moved off of the front page with the targeted killing of bin Laden in 2011, Obama maintained a sizable ground force in Afghanistan and dramatically expanded covert drone operations in multiple countries over the course of his two terms in office. Rather than ending, the war had, it appeared to his critics, simply slipped even further into the shadows.

Indeed, Donald Trump's comments at a ceremony honoring military veterans sought to reassure them that the negative war coverage they saw in the news wasn't an accurate reflection of reality, and that "it's all working out." The unpopular war he was referring to, however, wasn't the war on terror that many in his audience had fought in over the prior sixteen years in Iraq and Afghanistan, but rather the trade war that he had launched with China and other countries by introducing tariffs on various imported goods. The other war, the one in Afghanistan (which fifteen thousand US troops were still actively fighting at the time that he spoke), had receded so far from the mainstream conversation that Trump mentioned it only once, almost in passing.

While these remarks reflect widely varying politics and illuminate different points in the ongoing tide of the longest-running war in US history, they all rely on metaphors of vision and light—optics and occlusion—to signify this period of political instability and national uncertainty. The war they implicitly or explicitly reference is the ongoing war in Afghanistan, which began shortly after Cheney's remarks—part of the amorphous and ongoing war on terror. Where Cheney wants to prepare the public for the darkness and secrecy to come and Obama wants to signal a return to transparency and openness, Trump wants to shift public attention elsewhere entirely. All three men evoke images of conflict and darkness, but differ on whether it is beginning, ending, or simply not what it seems. What was this period of darkness?

The question elicits a range of possibilities, from the horrific nature of various historical events, to the disastrous, at times unethical, reactions of the leaders tasked with responding to them, to the acute polarization and political conflict that surrounded them. But it also emerges in part from the struggle to find sufficient images to represent and therefore comprehend historic events that were for many unbelievable. I do not mean that there were no images of these events—quite the opposite. The 9/11 attacks were widely viewed live on television, and the ensuing invasions of Iraq and Afghanistan were extensively covered on television through the military's embedded reporting program. Even events and policies hidden from the public and official media channels (the execution of Saddam Hussein, the torture program at Abu Ghraib, the drone program) produced a gusher of "leaked" images to fill the void. Nor am I referring specifically to places like the "Salt Pit" near Bagram Air Base or policies like "extraordinary rendition," which remained largely hidden from public view.[1]

Instead, our lack of sufficient images was a reflection of the political instability that the country faced at the dawn of a new historical period, one whose contours and dynamics were still occluded from view. Just as optical "seeing" often serves as a metaphor for cognitive understanding ("I see what you mean"), representing something in an image implies the ability to encapsulate and convey this understanding to another person. The period of darkness that emerged in the shadow of 9/11 was cast by our inability to understand or respond in a collective way to the events that unfolded as the war on terror began. In literal and metaphorical terms, we were unable to "picture" what was happening, incapable of "visualizing" a proper response. Instead of the "just" image we needed to shed light on our situation, we just had images—to reference Jean-Luc Godard's well-known turn of phrase.[2]

The optics of this dark period were, paradoxically, partially pushed forward into new territory by the same force that had driven their technological development for much of the previous two centuries: warfare. As scholars such as Paul Virilio, Friedrich Kittler, Lisa Parks, and Roger Stahl demonstrate, the connections between warfare and optical technology run deep, extending down into the classified realm of military research—where weapons development spins off into consumer technology and out into the popular culture, where in turn war and entertainment easily pass for one and the same spectacle.[3] Where the old model of two nation-states squaring off on the battlefield gave way to newer forms of urban guerilla combat and invisible insurgencies, modern military weaponry turned to the use of information and digital technology—cellular and satellite networks—to track and attack a disparate and disconnected enemy on a global battlefield. The complicity between the state, the military, and the entertainment industry that Tim Lenoir and Luke Caldwell powerfully outline demonstrates how all three collude in producing and profiting from the spectacle of war in society.[4] The blurring of legal and political lines between previously distinct entities and activities—like that between civilians and insurgents, or between military engagement and nation-building—produced a state of generalized confusion.[5] For Trump and his supporters, even the press and its coverage of these events became an "enemy of the American people".[6] These conflicts contributed to the larger impression that the United States was in the dark about who and where its enemies were—an impression mirrored in the increasingly "optic-less" and invisible form of warfare that the nation was waging.

But even as military strategy and government policy evolved and expanded to face these invisible threats, activists and protest groups took up similar technologies to expose and oppose these actions, a resistance that also drove the evolution of new forms of visual representation.[7] The opposition to these uses outlined above shaped their evolution as well. For as long as there have been optical recording technologies, there have been representations that sought to counter or contest their official, state-sanctioned use. Many of the filmmakers who pioneered these oppositional modes focused on altering the camera technology—effectively, hacking—as a way of altering or hacking the larger political and commercial framework in which it was developed.[8] This most recent period was no exception. As historical events forced the United States (in Cheney's eyes at least) to take unprecedented steps, these actions provoked reactions on the part of activists, artists, and everyday citizens. Like Cheney, these dissenting voices were forced to "work . . . the dark side," using "any means at our disposal" to articulate and disseminate their opposing viewpoints.

The result was an explosion of experimentation in nonfiction visual representation, one that combined older technologies like photography, film, and video with newer technologies like digital networks, social media, video games, and data visualization. These new, hybridized forms combined documentary aesthetics, political rhetoric, and digital technology. The same political instability that produced the war on terror brought each of the above fields into concert with one another—a development that produced a radical evolution in the technology we use to record and represent the world and ushered in the emergence of a new "worldview."[9] What brought us through the darkness was not a new technology, but rather a new way of seeing the world itself. This new way of seeing wasn't a product solely of the technology, the state, and the commercial institutions that developed it or the individuals who adopted and adapted it, but rather of the political conflicts and commitments that drove all three.

The birth of this new worldview is the focus of the text that follows. This book's title, *Where Truth Lies*, describes the space of transition from one worldview to another, a space where the disruption of what once seemed solid and trustworthy (or "true") forced the search for a new "truth" to replace what was lost. Rather than an absolute definition of "truth," it explores how the failure of existing representational paradigms to account for and describe a world that had suddenly been plunged into "darkness" compelled the creation of new paradigms on different representational ground. The space of experimentation between destruction and rebuilding—its own "ground zero" of sorts—is the space that this book explores. It emerges in the overlapping border of the three territories described above: politics, aesthetics, and technology. The competing worldviews it outlines are as much products of the dominant geopolitical regime (referred to in various contexts as "neoliberalism," "multinational capitalism," "vectoral capitalism," or "empire") as they are given forms of technological media or the aesthetic representations that they produce.[10] In this sense, the disruptions that occurred in the last decade reshuffled what Jacques Rancière refers to as the "distribution of the sensible." As Rancière puts it, "[T]he distribution of the sensible is the system of self-evident facts of sense perception that simultaneously discloses the existence of something in common and the delimitations that define the respective parts and positions within it."[11] The redefinition of where truth lies can be understood as a reorganizing of the "self-evident facts" and social relations that make up Rancière's distribution of the sensible. It is, in other words, a redistribution of the sensible.

Positioned at the cusp of film and digital media, this redistribution sits between what was clearly the dominant medium of the twentieth century (the moving image) and what is arguably becoming the dominant medium of the twenty-first century (digital technology and the Internet).[12] But the emergence of any new medium does not dictate the disappearance of its predecessors. Just as film did not eliminate its forerunners (still photography and print) and went on to survive the emergence of broadcast technologies like radio and television, moving images in general will survive and migrate over into new digital formats and distribution technologies.[13] And yet, this migration will inevitably change the nature of moving images, even as moving images equally impact the nature of these new media. Film, for example, may have survived and influenced the emergence of its moving-image counterpart, television, but it became nonetheless significantly different as a result. Similarly, film, television, and any number of other existing media continue to shape and be shaped by their integration with evolving digital technologies. The broad intersection of moving images and digital media thus represents the major media transition point occurring today.[14]

What follows focuses specifically on the interactions between one particular type of moving image—the political documentary—and the different ways it shapes and is shaped by digital media. Working at a transitional point in an ongoing historical process that is neither uniform nor monolithic, I explore a range of documentary and digital materials that bear the marks of their mutual influences to varying extents. Some of these texts are clearly recognizable as traditional, feature-length documentary films, and yet they nonetheless bear the imprint of digital technology in both form and content. Other works are "born" digital and lack any photographic, cinematic, or indexical trace, and yet they nonetheless fit within a broad documentary framework by virtue of their rhetorical contexts and political agendas. These hybrid media can be explained to a limited extent as *either* documentaries or new media, but they are best understood as some mixture of *both*.[15]

Political documentary and its long-standing connections to grassroots activism and independent media provide the ideal context for understanding much of the political activity taking place online and offline throughout this historical period.[16] At its root, documentary film collects information about the world, organizes this data into a socially meaningful form, and then presents this information to the public. In the context of political documentary, where abstract values like "justice" and "freedom" are called upon to address specific historical circumstances or social conditions, these aesthetic practices take on the role of rhetorical strategies intended to inform and persuade the public.[17] As an aesthetic practice, documentary instantiates or manifests this meaning through a variety of expressive codes, including framing, editing, exposure, sound, and so forth. When social conceptions of truth and meaning change over time, documentary aesthetics shape and are shaped by these wider ideological shifts.

This mode of aesthetic expression—organizing information into a socially meaningful form—also describes a great deal of the activity taking place in online environments. Indeed, a corporation like Google expresses this boldly in its mission statement, which proclaims that it wants to "organize the world's information." Although a digital form like a website or a database is not simply another form of documentary film, the same impulse to inform, educate, and persuade that gave rise to documentary film in previous periods shapes a great deal of the work being done through digital means. Moreover, documentary's long traditions of participant/independent media production, archival exploration, and social discourse/action all find correlates in interactive environments that seek to enable user-generated content, tagging, and social networking. Both expressions draw on the same social and political impulses, seek the same outcomes, and inform one another's execution even as they differ in the forms of meaning-making they undertake.

If part of the ideological disruption and redistribution that I describe lies in the interactions between these two forms of media, the rest of it lies in the specific political climate that emerged during the presidency of George W. Bush.[18] The extreme controversy generated first by his election in 2000 (the first of many close elections that reminded the electorate that "participation matters") and continuing through his handling of the 9/11 terrorist attacks ignited heated political debates throughout his term in office. These debates increasingly found expression *through* documentary means *in* online environments. Moreover, his administration's general antipathy toward the media and any type of transparency regarding its decisions and their consequences unsurprisingly fed a drive toward independent investigation and expression by both ends of the political spectrum. Indeed, Dick Cheney's assertion about the need to "work . . . the dark side" and "spend time in the shadows" also inadvertently came to characterize the administration's generally secretive approach. Consequently, the Bush administration became a natural target for a genre like documentary, steeped as it is in the Enlightenment ideal of transparency. This genre itself was in the midst of migrating to an environment that found its own mantra in Stewart Brand's well-known claim that "information wants to be free."

As the discrepancy between Cheney's policy of secrecy and Brand's mantra of transparency indicates, there is a discursive conflict between state politics and digital culture in the years after 9/11. These political debates and broader discussions about the Internet find themselves enmeshed in questions of information, democracy, human rights, truth, and the competing interests of the individual and society. From its inception, the war on terror was conceptualized both implicitly and explicitly as a war of information. Intelligence failures like the inability to predict the 9/11 attacks and the mistaken belief in weapons of mass destruction in Iraq were characterized as resulting from either a lack of information or the fabrication of information to lead the country into war. The controversial use of torture, on the other hand, is approached as a punishment of the body to access information in the mind.[19] Similarly, the spread of the Internet and networked culture are repeatedly evangelized as an information revolution ushering in an era of collective intelligence and universal access to information. As the torture debates were weighing out the relative importance of individual human rights vis-à-vis society's perceived need for safety, Internet communities were embracing Web 2.0 technologies to empower a cacophony of individual voices to be heard even as these individual voices were orchestrated into mass consensus (as a collectively authored site like Wikipedia illustrates).[20] Edward Snowden's shocking revelations that the US government was effectively spying on its citizens by covertly collecting information about the communications of its citizens sparked intense debate about privacy and state security.[21] As neocolonial political debates were waged about the need to bring democracy to the Middle East, the Internet was hailed as a democratizing source of information and positioned as the new public sphere that would return the demos to its rightful center at the heart of public life.

The meeting point between these discourses—between secrecy and the free flow of information, between darkness and exposure—is political documentary media. With its long tradition of alternative/activist media, political rhetoric, information dissemination, and collective spectatorship, documentary was ideally positioned to address both the prominent political questions that came to the fore during this period and the competing utopian and dystopian claims about the role of digital media in public life.

THE DIGITAL DOCUMENTARY

While the connections between each of these three areas of focus represent mature fields of research (film and digital media, documentary film and politics, politics and new media), the specific area this book addresses remains surprisingly underexplored. This is an odd omission for several reasons. First, significant moments in the development of documentary form and subject matter have repeatedly accompanied and coincided with developments in the technology used to produce and distribute these films. Discussions of the Direct Cinema and Cinema Verité movements of the 1960s almost as a rule begin with accounts of the development of production technologies such as improved 16mm film stock, wireless sync sound, and lighter, more mobile cameras.[22] And no less important to the Direct Cinema movement early on was the distribution these films received through television networks—another (relatively) new technology at the time. Both developments shaped how the films were made and seen by altering their content, form, and audience.[23]

Given the role technology played in prior moments, the emergence of digital video cameras, nonlinear editing, and streaming video on the Internet would seem to portend another period of rapid evolution. The explosion of streaming platforms such as Netflix and Amazon and the popularity and prominence of documentary content within these platforms partially demonstrate this.[24] But we should be wary of concluding that digital technology has simply ushered in a utopian better/faster/cheaper era for documentary film. Indeed, the move from the big screen of the theater at film festivals and organizational meetings to the small screen of the computer, smartphone, or flat-panel display in the home clearly involves a number of trade-offs for filmmakers, particularly in a genre where collective viewing is often fantasized as a form of collective action. Furthermore, a simple increase in image quality through camera technology, for example, has little effect on the quality of the final film as a film. Quoting Godard again, "There's no point in having sharp images when you have fuzzy ideas."[25] Nonetheless, these technological shifts did provide a number of benefits for documentary filmmakers, including expanded access to less expensive, higher-quality production equipment and a greater diversity of distribution options.[26]

Beyond improvements to image quality and distribution channels, the synthesis of digital technology and documentary film also produced more formally radical experimentation. This book will follow some of these experiments, seeking to locate them within the liminal space between the longer documentary tradition and the burgeoning applications of digital media. Doing so establishes the underexplored influence that documentary (its practices, ethics, and practitioners) had on digital media during a period when digital media was itself rapidly evolving and establishing a dominant presence in everyday life.

THE POLITICAL DOCUMENTARY

Exploring the impact of film and digital technology on documentary aesthetics implies a determinist connection between the two—a specter that haunts any study of the emergence of a new technology.[27] The influence of technology undoubtedly forms a key component of this evolution, but it is only one part of the equation. The radical formal innovation that this book connects with technology has also occurred during other historical moments. Documentary as the branch of filmmaking focused on the historical world has always been influenced by shifts in the society that it seeks to reflect and influence. Rather than a smooth development between history, aesthetics, and technology, documentary evolves at moments of historical rupture and social crisis: wars, economic depressions, and cultural revolutions. These underlying events and the political conflicts they expose are what drive the changes that I describe here. As historical events create specific challenges for society, filmmakers respond to these challenges by utilizing new and existing technologies to forge new rhetorical, formal, and aesthetic gestures.

Documentary as a genre is a notoriously capacious concept that covers a heterogeneous variety of rhetorical modes and poetic registers, from state propaganda to cultural ethnography to autobiographical expression. However, certain historical periods appear to have pushed the integration of new technology into the wider documentary lexicon. In particular, three prior historical moments reveal an evolution similar to the digital evolution I am addressing: (1) the 1930s and 1940s; (2) the late 1960s and early 1970s; and (3) the late 1980s and early 1990s. At each of these points, documentary filmmakers utilized new forms of technology to respond to specific social and political crises. In the 1930s and 1940s, film-makers such as John Grierson, Pare Lorentz, Frank Capra, and Leni Riefenstahl utilized new sound technologies to create state-sponsored films that addressed social issues related to the Great Depression and World War II. Their films utilized what would eventually be termed "voice of God" narration to articulate the aims and nationalist endeavors of the state. As previously mentioned, in the 1960s and 1970s, American Direct Cinema filmmakers and other collectives like Newsreel utilized newly available 16mm sync sound equipment, faster film stock, and lighter cameras to document the rise of the American counterculture and stimulate resistance to the war in Vietnam. This countercultural ethos was once again mirrored in the form of their films, all of which shunned voice-of-God narration in favor of observation and participant interviews. And finally, the work of filmmakers such as Jill Godmilow, Rea Tajiri, Jenny Livingston, Marlon Fuentes, and Marlon Riggs responded in the 1980s and 1990s to the culture war issues of the HIV/AIDS crisis, Reagan-era cutbacks of social welfare programs, the defunding of the arts, and other issues by turning to video to create deeply personal films meant to reflect the issues of specific subgroups. These performative films sought to bring what many had deemed the obscene "on/scene" by exploring the experiences and identities of these groups in a visible, mainstream form of media expression.[28]

These previous periods also demonstrate significant developments in documentary's rhetorical approach to what Jerry Kuehl calls its "truth claims" about the historical world.[29] As different groups turn to documentary to further their social and political aims, they bring with them very different theoretical and philosophical influences, which in turn produce different rhetorical strategies for speaking about the world. As Jonathan Kahana points out, voice-of-God documentary in the 1930s and 1940s, which sought to provide governmental transparency to the populace, originated in the pragmatist philosophy of thinkers like John Dewey,

Walter Lippmann, and A. D. Lindsay. For these theorists and the filmmakers they influenced, the general public needed to understand the true nature of the complicated workings of large government bureaucracies; hence the need for films that performed this edifying function.[30] As historians of activist collectives like Newsreel have pointed out, theorists such as Herbert Marcuse influenced young activists to pursue independent representations of truth in opposition to the ideological truths of the state and mainstream media. In other words, countercultural filmmakers were offering the people's truth to counter the dominant state truth.[31] Finally, films in the 1980s and 1990s reflect the "post" systems of thought (postcolonialism, postmodernism, post-Fordism) that deconstructed master narratives of truth, giving way to smaller-scale truth claims regarding the legitimacy of alternative identity formulations and a conscious utilization of aesthetic form as an expressive tool.

It is worth pointing out that the films that I am highlighting here and the innovative formal characteristics they exhibited do not necessarily represent the dominant or most popular styles of the time. Instead, they demonstrate relatively new modes of expression that came to characterize a particular political or ideological position within their respective contexts. This stylistic innovation emerged from creative uses of new technology—a move driven out of the desire to push political expression in new directions. But these formally inventive experiments emerged alongside plenty of more conventional work. Similarly, the characteristics that I will outline in the following chapters are not necessarily the dominant mode or style of post-9/11 documentary. Many remain one-off experiments that failed to generate any significant trend or imitation. Nonetheless, all are noteworthy because they managed to establish something new within the political conversation where they appeared. Many in fact achieved this influence because they originated from and came to represent previously marginalized groups. This was particularly the case with the formally innovative films that appeared in the late 1980s and early 1990s.

THE AESTHETICS OF POLITICS

The performative documentary of the late 1980s and early 1990s deserves extended discussion, not only because it directly precedes the 9/11 political context but also because it was a point at which aesthetics self-reflexively emerge as a fundamental tool (one is tempted to say "weapon") for politically motivated work. Whereas the two prior periods utilized a particular aesthetic form to express and instantiate the political positions they espoused, for both, questions of aesthetics were an unacknowledged, almost regrettable, component of documentary expression. Indeed, one way to characterize the move from the institutionally driven, voice-of-God films to the Direct Cinema and Cinema Verité films is as a conscious move to limit artifice in favor of directly exposing or observing what was taking place before the camera.

For groundbreaking political filmmakers in the 1980s and 1990s, however, aesthetic form was consciously embraced as the fundamental ground on which the political battle was to be waged. As before, changes in production and distribution technologies (the emergence of consumer-grade video cameras and public access on expanding cable networks) emerged alongside a broader epistemological shift about the problematic nature of different forms of representation. Scholars and practitioners across fields such as history and anthropology as well as the arts and humanities began to question the ethics and accuracy involved in their work. As Hayden White demonstrated, clear distinctions between form and content blur to the degree to which meaning becomes a product of the particular narrative form that a given texts adopts.[32] Ethnographic filmmakers like David MacDougall, building on the work of anthropologists like Clifford Geertz, began self-consciously inserting themselves into their stories, implicitly undermining the objective authority once taken for granted by the camera/observer. In the work of practitioners and theorists such as Trinh T. Minh-ha, aesthetic form became central to documentary's legitimacy as a mode of expression rather than a regrettable drawback to be avoided at all costs.[33]

These insights shifted documentary expression past its "talking head" experts and fly-on-the-wall observations to a more experimental, expressive mode of representation that consciously utilized and maximized aesthetic form rather than minimizing its intrusions. Even a more mainstream filmmaker like Michael Moore, whose 1989 *Roger & Me* falls directly within the performative era I am describing here, filters his populist politics through an individual, autobiographical framework, tracking the impact of capitalism on his hometown of Flint, Michigan, and appearing on-screen as the dominant persona of his films. During this period, documentary film, fine art, experimental film, video art, and grassroots activist practices all began to cross-pollinate.[34] This created a level of fluid hybridity between forms and modes that would reemerge in the post-9/11 period.

But this prior period of political conflict and formal experimentation also provided a paradoxical legacy for documentary filmmakers and independent activists responding to the war on terror. On one hand, in 9/11 and the war on terror they were confronted for the first time since the Vietnam War with an acute crisis of national magnitude. The scale of the conflict and the urgency of the issues at stake created the impulse to speak to the widest possible audience about divisive political issues. On the other hand, they were utilizing a medium that seemed to have lost its rhetorical power to mobilize a mass population, one less willing than ever to take to the streets and march on anything.[35] Indeed, the formal turn in documentary filmmaking during the 1980s and 1990s brought a conscious rejection of a single, universal political truth and a recognition that documentary aesthetics were part of the content of the form. This would seem to leave little space for either formal innovation or political engagement. And yet, the period after the 2000 presidential election and 9/11 ignited an era of widespread political engagement that was reflected in a surge of politically focused documentary film production.[36]

The convergence of political and technological forces cast all of the received categories up for grabs, creating an opening for this new period of formal evolution. Rather than a "universal truth," much of this experimentation was directed toward exposing a particular injustice against a common enemy. For example, many political documentary films began to levy their truth claims over and against those offered by the mainstream media, as Charles Musser has argued.[37] That is, instead of purporting to present a state truth or an antistate truth, as films in the 1930s/1940s and 1960s/1970s did (or, in the case of 1980s documentary, an anti-normative truth), documentary films after 2000 counterpose their truth claims against an increasingly polarized and politically distrusted "media" truth. (For the political right, the object of this distrust is typically the *New York Times*, while for the left, the role is filled by Fox News.) This rejection of mainstream news sources only increased throughout this period, reaching a crescendo in the divisive relationship between Donald Trump and what he refers to as the "fake" mainstream news.

Moreover, in the last thirty years, documentary films that address political issues seem to have realized the futility of speaking to those beyond an audience of sympathetic viewers. While this audience has often been a specific constituency, the polarized political environment in the wake of the contested 2000 presidential election simplified and solidified these constituencies into two major wings. (To paraphrase George Bush's own terms, "You're either with us or against us.") This divisiveness enlarged the potential constituency for any given film to roughly half of the population. Thus, for the first time since Vietnam, documentary films had the freedom to speak *to* specific groups that were already inclined to believe the truth claims they presented *about* issues that affected everyone. This expanded audience provided the perfect environment for the renaissance of political documentary and documentary experimentation produced in the first decades of the twenty-first century.[38]

The historical legacy of documentary film as a viable form of political action and mobilization accounts for its reemergence on the national scene. As an activist form of independent media, documentary provided an alternative media model for a number of activists organizing in various digital environments. The generalized distrust of the mainstream media that characterizes our current moment partially explains the resurgence of documentary's popularity, but we should also attend to the incorporation of other alternative forms of participant media, from weblogs and podcasts to Twitter feeds and wikis. The same impulse that gave rise to political documentary films in prior moments is now a driving force in the way these other, newer forms of digital media are deployed in current social and political battles.

Each of the following chapters offers a comparative analysis of two different media objects or texts situated at some point along the spectrum between documentary film and digital media. I focus exclusively on the historical period that begins with the dot-com crash of 1999, the 2000 election, and the 9/11 terrorist attacks and ends with the election of Donald Trump. However, I draw on prior moments as they inform the central texts I consider. Given that many of the political issues that arose during the administration of George W. Bush scarcely subsided with the return to power of the Democratic Party and reemerged with the Trump administration, my concluding chapter follows this formal evolution into the era of "posttruth politics" and "fake news" that characterized the 2016 election.

Chapter 2 explores the impact of nonlinear digital editing and compositing programs on traditional documentary film by presenting a close reading of the recent work of Errol Morris—particularly his two films about war, politics, and technology: *Standard Operating Procedure* and *The Fog of War*. In the arc of Morris's career, the two films present something of a paradox. On one hand, both deploy all of the standard tropes of his by-now signature style: Interrotron-enforced eye contact, richly staged reenactment, plentiful archival material, and visually dense montage sequences set to hypnotic music. Moreover, both films deal directly with questions of war, the military, and the mediated, moving-image representations we have of both. And yet these films respectively represent the high and low points of his career. Though somewhat controversial with critics for its treatment of Robert McNamara, *The Fog of War* generated nearly $21 million at the box office and garnered Morris the Academy Award for Best Documentary. Going into *Standard Operating Procedure*, it seemed as though his films had finally received the mainstream attention and studio backing many felt he had always deserved. The success, however, was short-lived as the film debuted to even greater controversy and went on to fail commercially as well (earning just over $300,000 total). Leaving aside why one film succeeded where the other failed, my reading claims that the controversial subject matter both explore allows Morris to address a larger point about the capability of media representation. I argue that this collage style is now the product of a certain form of *database aesthetic* in which the elements of the historical record (the archive) act as discrete elements that can be mixed and remixed depending upon the particular discursive context one wishes to construct.[39]

The irony inherent in his critique of media representation is that at this stage in his career Morris himself can almost be considered a multimedia artist. Both of these films were extensively expanded in companion books, and Morris contributes material regularly to a blog for the *New York Times* as well as to his webpage,

ErrolMorris.com. That is, as Morris's films critique the omnipresence of representational recording technology and demonstrate the extent to which it can lead us to disastrous conclusions, his other outside activities contribute to the saturation he criticizes. These extracinematic materials also demonstrate the beginnings of a symbiosis between film and other forms of media.

But radical experimentation was also taking place elsewhere. If Errol Morris demonstrates the influence of digital technology and documentary's encroachment into other forms of media, Robert Greenwald's 2003–4 film *Uncovered: The War in Iraq* and its relationship to the political advocacy group MoveOn.org reveals a full-fledged interdependence between the two. As grassroots political-organizing groups moved into online spaces in the early days of the web, they pioneered the use of streaming video as a recruitment tool. Indeed, a quick glance at political-action websites from the period ranging from those on the left (Reprieve.org, Witness.org) to those on the right (RightMarch.com, TeaPartyPatriots.org) reveals a universal reliance on streaming video footage to articulate a group's message and document its past action. The use of video in these contexts demonstrates a crucial synergy between newly available online technologies and the century-old documentary tradition. As the social web slowly emerged and took shape in the early 2000s, online advocacy groups were experimenting with community building and online organizing. Documentary's ability to marshal evidence with argument to present a call to action found a natural home on websites capable of providing an immediate outlet for the impulse.

Chapter 3 analyzes one of the key historical moments in the remediation of nonfiction video and online political organizing: the 2004 collaboration of director Robert Greenwald (*Outfoxed* and *Walmart: The High Cost of Low Prices*) and the political action powerhouse MoveOn.org. This chapter provides both a historical account of the release of *Outfoxed* and an analysis of the extent to which MoveOn and Greenwald relied on one another to achieve their specific goals. Utilizing this cross-pollination as a case study, I analyze the mutual synergies between these two forms of media in order to demonstrate the necessity of each for the other as well as to interrogate the extent to which both forms still rely on real-world action to achieve their ends. Even as political action and its attendant images continue to move online, the goal remains to move people toward direct action in the real world. The efficacy of this new form of digital activism and its comparison with other forms of activism also form part of the story that unfolds here.

Whereas chapter 2 presents a traditional documentary film that bears some of the imprint of digital logic and chapter 3 explores the use of documentary images in online spaces, chapter 4 presents wholly digital media influenced by a documentary logic. It offers a comparative analysis of two online, interactive video games with clear political overtones. The first is the *Gone Gitmo* project, which attempted to re-create the real but inaccessible Guantánamo Bay prison complex in Second Life, a virtual but accessible online environment. The project uses various "documentary" sources for its re-creation and seeks to raise awareness about the political issues involved in the prison. The second is the *America's Army* video game, which, since 2002, has served as a recruiting tool for the US Army. The game offers players the chance to participate with other players in missions that simulate battles in Iraq and elsewhere. Although both games completely forsake the photographic indexicality that is documentary's tie to the historical world, both games utilize the documentary impulse to intervene in and motivate the individual to act in the real world. Both texts engage in a give-and-take exchange with reality in ways that mirror earlier documentary film efforts, and yet, the mechanisms that drive this exchange are clearly different. Such tactics were utilized by the state in propaganda training and recruitment films like the *Why We Fight* series in World War II and by individuals in both observational and performative documentaries in the 1970s and 1980s. As the cultural theorist Johan Huizinga points out, the move from spectator to player involves the participant in a methectic rather than mimetic relationship to the representation. This move gives *Gone Gitmo* and *America's Army* a powerful form of interpellation that engages participants in the new forms of subjectivity that both seek to achieve.[40]

Chapter 5 looks at the use of data visualization in government transparency initiatives during the first years of the Barack Obama administration. While Obama attempted to use twenty-first-century tools to create what he called "the most open administration ever," his efforts were ironically sidestepped by the unprecedented information releases of the anarchist/activist group WikiLeaks. At the same time that Obama was seeking to digitize and visualize government records, WikiLeaks was actively challenging the barriers between public and private entirely. Even as WikiLeaks reached for a series of optically driven Enlightenment metaphors to characterize its objectives (light, truth, sunshine, transparency), both utilized non-optical media to achieve this. Moving from the documentary image to digital data in an aestheticized form, chapter 5 provides a historical look back into the history of the transparency debates while also looking forward to the shape these debates will play in coming decades.

Chapter 6 considers two of the primary by-products of the bitterly fought 2016 presidential election and the emergence of what many have characterized as the "posttruth" political style of its surprise winner, Donald Trump. The first is the role being played by conspiracy theory and paranoid speculation, both of which fueled Trump's entrée into national politics and clouded his ascension to the White House. Prior to the election, many of Trump's critics highlighted his relationship with dubious theories and fringe political figures, including Alex Jones of InfoWars, as evidence of his dangerous disregard for the truth. Exhibit

A was Trump's promotion of "birtherism" (the theory that Barack Obama was not born in the United States) as a race-baiting means of gaining attention and undermining the legitimacy of the sitting president. After the election, however, as news began to circulate of possible Russian interference, election hacking, and secret dossiers, many of these critics fell victim to the same forms of conspiratorial speculation that just months earlier had seemed at best delusional, and at worst, irresponsible. As a form of independent speculation and alternative-media creation that seeks to question power and challenge existing narrative frameworks, conspiracy media and the theories it promotes offer a version of nonfiction media not entirely divorced from documentary film. And yet, the mode of paranoia and suspicion in which they operate forms a corrosive, adversarial context in which to debate and engage political ideas.

The second object this chapter considers is the emergence of a new breed of independent news outlet that I refer to as the "midstream media." Positioned between the mainstream media and the more fringe figures like Alex Jones, these midtier players exemplify the new breed of news organization that played a pivotal in the election and the so-called fake news debates that it produced. Steven Bannon's Brietbart News Network and Glen Greenwald's news organization The Intercept offer good examples of the midstream market more broadly. Ideologically, the two sites could not be further separated, and yet they bear a striking material resemblance to one another: both are independently funded by a select group of wealthy individuals, both occupy a position somewhere between mainstream media and the polarized extremes of independent media, and both critique the failure of existing media to hold governments accountable and provide truth to the public

While both of these developments (the rise of conspiracy media and the more polarized midstream media) seemed to emerge in the context of the 2016 election as endemic qualities of Trump and his unique political brand, this chapter demonstrates that both are the result of longer-standing trends in American political culture. Indeed, both emerge and grow as a result of the same forces that shaped all of the different objects that this book considers: a rapidly evolving media environment ever more suffused by digital technology, and a highly charged political environment that emerged in the wake of 9/11. In this sense, the chapter continues the formal construction and political/technological focus of the earlier chapters while providing a conclusion to the argument as a whole.

Taken together, these chapters explore the emergence of the "digital documentary" by placing equal weight on both sides of the term, arguing that documentary informs the digital as surely as the digital informs documentary. As "data" becomes the central lens through which we view ourselves and the world around us, existing nonfiction practices such as documentary film will be the primary media that shape how digital media impact and express our individual worldviews.[41] In seeking legitimacy as a mode of expression, digital media drew on the precedent set by documentary, and in grappling with the impact of digital media on our lives, documentary began to express the different fears and fascinations surrounding the transformation. If the first decades of the twenty-first century did indeed cast us into the dark, as our various political leaders would have us believe, it was a combination of both digital culture and documentary film that brought us back into the light—an evolution of forms and practices that worked to show us "where truth lies" for the century to come.

2

"We See What We Want to Believe"

Archival Logic and Database Aesthetics in the War Films of Errol Morris

Nothing is more frightening than a labyrinth that has no center.
—G. K. CHESTERTON

INTRODUCTION

In May of 2000, in connection with the premiere of his television series *First Person* on the Bravo network, documentary filmmaker Errol Morris launched his first home page on the World Wide Web at www.errolmorris.com. Initially the site greeted visitors with the Chesterton quote above—a wry commentary on the rhizomatic, decentralized structure of the web. But beyond this, it offered little more than the standard webpage info (biography, filmography, interviews, etc.— what the site would later link to as the "BORING stuff"). After a few months, however, Morris published a black-and-white image of a horse's skull with crosses over the eyes next to the following list:

Why It Makes Sense to Beat a Dead Horse

1. Sets an example for other horses
2. Aerobic workout
3. Horse might not be dead yet
4. Tenderizes the meat
5. Horse is unable to fight back
6. Makes you feel good[1]

This list was one of several the site would feature over the coming months and years. (Others include "Why It Makes Sense to Bite the Hand That Feeds You" and "Why It Makes Sense to Wear an Albatross around Your Neck"). It represents

18

Morris's first attempt at creating original content for his newly adopted medium and foreshadows something of the random, ironic tone that he would develop further on the site over the next decade. As it stands today, the site is a teeming labyrinth all its own, with content drawn randomly and in connection with his many film projects, commercials, books, blogs, tweets, and other media that the director now uses to explore his selected topics of interest. Far from a simple website promoting his moving-image work, the site is a full-fledged creative production of its own, and one of the more interesting utilizations of the Internet by a filmmaker to connect and expand upon a multimedia body of work.

As the website and its collection of content demonstrate, Morris occupies a unique position in the field of documentary film. On one hand, he ranks among the more prominent American documentary filmmakers, standing alongside other mainstream directors like Michael Moore and Ken Burns. On the other hand, Morris has embraced digital technology head-on, incorporating it formally and thematically into his cinematic work and as a new medium in its own right through his website, blog, and social media accounts. Given his reputation as a director willing to take on such abstract topics as truth and human perception, Morris's work also became increasingly relevant (and controversial) in a period marked by extreme political polarization and overt ideological confrontation in the United States. Throughout his career, Morris's films have always been structured around a basic tension between subjective fallibility and objective truth, or, put differently, between individual delusion and social history.[2] He has a well-established body of work dealing with both the intersection between eccentric personalities with unique perspectives (*Vernon, Florida* [1981], *A Brief History of Time* [1991], *Fast, Cheap & Out of Control* [1997]) and human access to the past via memory and evidence (*The Thin Blue Line* [1988], *Mr. Death* [1999]). After 9/11, Morris's projects expanded this focus to include a more direct interrogation of the role of specific forms of media in altering or enabling our access to events in the world. In this category we could include *Tabloid* (2010), *The Unknown Known* (2013), and both *The Fog of War* (2003) and *Standard Operating Procedure* (2008).

Morris's first two film projects after 9/11, *The Fog of War* and *Standard Operation Procedure,* demonstrate an acute concern with the impact of digital technology on politics and warfare and a deep integration of technology into the text of the film. Beyond simply incorporating CGI and other digital effects into their production, these two films demonstrate a willingness to interrogate the widespread influence of such technologies on individuals and their perception of the world around them. The dense collage of archival material that confronts us in *The Fog of War,* for example, inherently encapsulates and interrogates the archival logic that surrounds much of the drive behind the Internet today.[3] In *Standard Operating Procedure* the focus turns to a specific form of media—digital photography—directly addressing the plasticity of meaning that a database of digital imagery affords. This combination of factors makes Morris's output (both online and on-screen) the ideal object for charting the convergence of these forms in the period after 9/11. These films are *about* digital media as much as they are products *of* digital media.

As the political controversies of the twenty-first century succeeded one another with astonishing rapidity (the 2000 US presidential election, the 9/11 terrorist attacks, the wars in Afghanistan and Iraq, the open-ended "war on terror," the Patriot Act, and the Guantánamo Bay prison camp, to name a few), such issues were increasingly presented and debated in a newly fragmented media landscape divided between old and new media. Like the web, American politics increasingly *became* a confusing labyrinth of information and obfuscation, a maze without a center. Thus, the principles that had long structured Morris's films increasingly seemed to structure American political discourse as well. Dealing with former Secretary of Defense Robert McNamara and with the controversial Abu Ghraib photographs, respectively, *The Fog of War* and *Standard Operating Procedure* are Morris's most overtly political works to date. Unlike his previous works, which uncovered the more obscure corners of the world, both films address people and events that had widespread social impact, and both focus on war and the technological media used to wage and represent it. In doing so, these films and their multimedia offshoots enter the labyrinth of images that shape our collective view into past and present, thereby offering an entry point into the evolution of technology, politics, and aesthetics during the decade after 9/11.

THE FOG OF WAR'S TWIN LOGICS

The Fog of War is structured loosely around eleven lessons drawn from the life of Robert McNamara. As James Blight and janet Lang make clear in the eponymous book that accompanied the film, these lessons are themselves the product of a series of conversations and conferences that McNamara participated in along with other leaders via the Wilson Institute's Critical Oral History conference series. Initiated by Blight, the project brought together former policymakers and academic experts to debate the events and records that make up our collective understanding of the past. Many of these reflections had previously been collected in a volume called *Wilson's Ghost*, coauthored by McNamara and Blight.[4] In the books, these reflections take the form of a series of aphorisms drawn from McNamara's direct participation in key historical events like the Cuban Missile Crisis, the Vietnam War, and to a lesser extent World War II. In essence, they are positioned as history lessons, not in the sense that they hope to teach us facts about the past, but rather that the past itself is offering us insight into how to do things differently in the future.

Thus, much of the film's formal structure and the conclusions it draws regarding its subject existed well before McNamara ever stepped in front of Morris's Interrotron.[5] But of course, the film itself is much more than a moving-image interpretation of thoughts put together elsewhere. That is, in the process of translating this material to the screen, Morris adds his own interpretation of the lessons these events can teach us and his own view on the perspectives of his subject. Among these lessons are two that form the core of the film's critique: "Lesson 2: Rationality Will Not Save Us," which the film uses to critique the use of computer-driven logic and statistical control in warfare; and "Lesson 7: Belief and Seeing," through which the film interrogates the relationship between images and the events they document and communicate.[6] Both lessons perform the double function of depicting McNamara's recollections on-screen while at the same time setting up the film's larger conclusions about our own computational and photographic approaches to the past and, indeed, to reality itself.

McNamara is in many ways an ideal figure to explore the connections between media, the rise of computational logic and control, and politics and warfare. He is most associated with being the early architect of the Vietnam War, an ambitious bureaucrat who was appointed by President John F. Kennedy to oversee and overhaul the sprawling Department of Defense. He eventually led it down the path of its most disastrous military endeavor. In a sense, he was neither a media figure nor a computer scientist, but his time in politics arrived when computation and media were becoming essential elements of both. His tenure as the head of the Pentagon witnessed one of the most ambitious integrations of warfare and computation to date, and his Department of Defense was one of biggest investors in the early stages of computer networking and remote command and control.[7] As the United States launched its war on Afghanistan in 2001 and word of the first drones and other technologized weaponry began to dominate the news cycle, these issues were back in the headlines. It is worth recalling that these were connections that began when McNamara was at the head of the Department of Defense. These are connections that Morris was certainly aware of, and they are themes that permeate the film.

To get a sense of these larger conclusions, we need look no further than the opening of the film. The first footage we see is a grainy, black-and-white television recording of a young McNamara standing behind a podium adjusting the height of a chart and asking his audience if this is "a reasonable height for people to see." The camera then cuts to McNamara at the podium, where he states: "Earlier to-night . . . let me first ask the TV 'Are you ready? . . . all set?'" Just as he is about to begin again, the film cuts to the opening credits. Intercut with the credits and set to Phillip Glass's score are more grainy, archival shots of soldiers on a ship looking out at the horizon using various devices (binoculars, sonar equipment, maps, and charts) and apparently preparing for a battle of some sort.

Taken together, these two brief moments hint at the primary themes in the film. We are introduced, via the news footage, to McNamara not just as the film's main subject and sole interviewee but further as someone who is media savvy and thoroughly controls the message he is about to send. This is a message, moreover, that will be delivered with the aid of charts and graphs, delivered in a manner that's "reasonable" to the audience. Reducing the impact of what he wants to say for those assembled in the room with him at the time matters less than making sure that the "TV" is ready. The film's opening, an ironic "behind the scenes" beginning from the past, also serves as an indicator and a reminder of the manipulated nature of the media through which such messages are transmitted. Lest we miss it, the closing of the opening credits gives way to the following exchange between McNamara and Morris:

McNamara: Let me hear your voice level so I can know if it's the same.

Morris: *[off-screen]*. How's my voice level?

McNamara: Fine. Now I remember exactly the sentence that I left off on. I remember how it started, and I was cut off in the middle, but you can go back and fix it up somehow. I don't want to go back and introduce the sentence because I know exactly what I want to say.

Morris: Go ahead.

McNamara: Okay. Any military commander . . .[8]

As in his archival appearance before the cameras, McNamara is once again fully in control of his message, to the extent that he suggests how Morris should eventually edit the film by "fixing it up somehow." Rather than take this advice, Morris instead chooses to include it, reminding us once again that such messages are shaped and framed not just by those who send them but also by the media that transmit them.

The footage in between these two clips is no less significant. As described, it consists of various soldiers on a battleship studying their environment and preparing to act on their observations. Although presented only in brief segments lasting no more than a few seconds each, they all depict what must be a very routine set of events in a hostile environment. A situation is observed via optical, infrared, and topographic means (binoculars, sonar, and maps, respectively) in order to determine the proper response. Once a decision has been made, the information is communicated and a course of action is set. This, of course, is no different than what most of us do in every waking moment as we observe and respond to our environments, but in this case the stakes are far higher; given the presence of massive cannons and the assembly of bombs and other munitions, these actions and perceptions become a matter of life and death.

Taken together, these reminders of the mediated nature of media and the archival footage of preparing for battle offer the viewer a stern warning about the information we use to reach our own conclusions and determine our actions as we take in the flow of information from the media that surround us. We should be on guard, it seems, not just against the potentially flawed and mediated messages we receive, but also against the conclusions we make and the actions we take based on those messages. This point is further reiterated and explored in the two "core" lessons from the film.

"LESSON 2: RATIONALITY WILL NOT SAVE US"

Throughout the book version of *The Fog of War*, as well as in the other written material by McNamara and Blight, the aphorism that "rationality will not save us" forms the backbone of their reflections on the Cuban Missile Crisis.[9] This is the point that McNamara puts forth in the film as well. Throughout the documents collected in the text, some of which are excerpted in the film, the authors paint a picture of a world standing at the brink of a nuclear war that is narrowly averted at the last minute by one factor: luck. As McNamara puts it in the film:

> I want to say, and this is very important: at the end we lucked out! It was luck that prevented nuclear war. We came that close to nuclear war at the end. [*Gestures by bringing thumb and forefinger together until they almost touch.*] Rational individuals: Kennedy was rational; Khrushchev was rational; Castro was rational. Rational individuals came that close to the total destruction of their societies. And that danger exists today. The major lesson of the Cuban missile crisis is this: the indefinite combination of human fallibility and nuclear weapons will destroy nations.[10]

Thus, for McNamara and Blight, the danger posed by nuclear weapons lies in the irreversibility of a single bad decision in the face of a conflict like the one in which Kennedy, Khrushchev, and Castro found themselves in October 1962. Even rational leaders such as these can make a reasonable choice based on faulty information and incorrect assumptions that will lead to disastrous consequences. Surely this seems accurate, and nothing in the film works to contradict it.

In their dismissal of the ability of rationality to solve such problems, both Blight and McNamara leave oddly unexplored the role that rationality plays in creating them. That is, by pointing to rationality's failure at a key historical moment, they miss the extent to which it was responsible for producing this moment in the first place. This lesson is not lost on the film. *The Fog of War* spends a good deal of time visually exploring the role that instrumental rationality played in creating McNamara's own perspective. This critique arises subtly from the structure of the film's visual materials. Shortly after the opening sequences examined above, the film introduces this theme through archival footage from a *CBS Reports* segment entitled "McNamara and the Pentagon."[11] As observational footage rolls of McNamara scribbling down graphs and percentages for a group, a voice-of-God narrator introduces him with the following description:

> This is the secretary of defense of the United States, Robert McNamara. His department absorbs 10 percent of the national income of this country, and over half of every tax dollar. His job has been called the toughest in Washington, and McNamara is the most controversial figure to ever hold that job. Walter Lippmann

FIGURE 2.1. A soldier scans the horizon in *The Fog of War*.

calls him not only the best secretary of defense but the first one who ever asserted civilian control over the military. His critics call him a con man. An IBM machine with legs. An arrogant dictator.[12]

The nomination of McNamara as an "IBM machine with legs" is one the film underscores throughout via other archival materials and reenactments. For example, when he discusses his biography and the events that led to his involvement in World War II, McNamara describes his role in creating the "US Army Air Corps Statistical Control School" in 1942, a post that led directly to his commission in 1943 as a lieutenant colonel in the Air Corps, overseeing logistics and success rates in the air campaigns over Europe and Japan. McNamara hints here that one of his great achievements was the insistence that the school take the punch cards on which the military had collected data on every soldier and run them through the IBM sorting machine for criteria like "age, education, accomplishments, etc." "We were looking for the best and the brightest. The best brains, the best capacity to lead, the best judgment."[13] McNamara thus positions his ability to act in a rational manner using logic and statistics as among the key factors in his success both at Harvard and in the military afterward.

But the film's image track throughout this segment is telling. In addition to the interview footage of McNamara, the film oscillates between archival footage of animated charts with titles such as "Analysis of Striking Power in Heavy Bombers in ETO" and reenacted footage of punch cards sliding through an IBM Hollerith tabulating machine—the very same IBM machine to which McNamara's critics compared him. The footage of the Hollerith foregrounds this earlier criticism and initiates a chain of associations that that film directly connects back to its subject. Developed for the 1890 census, such machines have long been synonymous with statistical information and population control.[14] Moreover, during the period of the film's production, a minor controversy erupted regarding the role of IBM's complicity with the Nazis and the role of the Hollerith in the German war machine.[15] By invoking the comparison between McNamara's own thought process and the mechanized efficiency of this early computer, the film establishes a visual metaphor that unites computational logic and human rationality with inhuman aggression and destruction—a theme reiterated each time this same footage reappears. Even as McNamara points to the importance of rational decision-making, the film pairs this form of rationality with acts of violence and aggression. Most damningly, as McNamara states that he wanted people with "the best judgment," the image track cuts to footage of bombs falling from a plane.

McNamara, of course, was not alone in introducing rationality, computers, and statistics into the perfection of warfare. The historical role of other academics such as Alan Turing and Norbert Wiener in the creation of encryption and targeting systems for the military has been well established.[16] Nor is he the only one to paint these activities in a positive, patriotic light—as having had a beneficial impact both on the war effort and on society in general. After heading the military's Office of Scientific Research and Development, Vannevar Bush famously lamented the loss of a common research goal that the end of the war would bring and called on scientists to collaborate in creating tools that would enable them to share and communicate more effectively during peacetime. One such solution was a tool called the Memex, based on a technology that many see as an early model for hypertext and the Internet.[17] Others, however, rethought the ethics of applying science to warfare. Wiener, for example, even went so far as to forgo any type of military funding for his postwar research.[18]

Unlike his academic counterparts who took part in the war effort, McNamara declined to return to academia and opted instead to put his newly perfected optimization and rationalization procedures to bear on production and design in private industry for the Ford Motor Company. Here, McNamara describes once again the importance of personality testing (accompanied again by shots of the Hollerith) and explains how he set up a marketing office to "get the data" about who was purchasing cars. He also describes commissioning research on accident statistics to understand how to manufacture safer vehicles. The image track cuts between various charts and graphs, again visualizing McNamara's approach to solving problems. The problem, he states, was "packaging," or the materials that surround and secure the driver in the car. This determination led him, with the help of scientists at Cornell, to research how the human body could be better protected by dropping human skulls wrapped in various materials down the stairwells of the school's dormitories.

Here, the film cuts to what Morris describes in an interview as his favorite shot of the film.[19] As he tells Terry Gross on NPR's *Fresh Air*, "Whenever I hear a story, particularly if it's a good story, an image comes immediately to mind and it becomes very hard to resist the temptation to shoot those images. . . . [P]art of *The Fog of War* is a story of dropping things from the sky, bombing if you like. . . . But this is an instance where dropping things actually produces good rather than evil."[20] The slow-motion shot Morris produced to illustrate McNamara's anecdote is thus in part one of the redemptory moments in the film for McNamara and the rational approach that he expounds throughout.

As with all of Morris's reenactments, there is something here that exceeds the image's purported meaning. The image of a human skull falling in slow motion through space and eventually smashing into pieces at the bottom of a stairwell opens itself to any number of readings beyond simply illustrating McNamara's story of dropping things for "good rather than evil." On one hand, the skull has long been the symbol of death and mortality, a reading compounded by the frailty it demonstrates in coming apart as it hits the stone surface below. Thus, we are reminded of the true cost of calamities like auto accidents and high-tech warfare. On the other hand, the skull is itself the "packaging" for the human brain, the seat of the thought and rationality that the film reminds us again and again will not save us. Its destruction in this sense speaks to its fragility in the face of "dropping things." Regardless of which reading we choose, the image nonetheless presents a damning indictment of the application of rationality to human aggression that McNamara celebrates throughout the film. Again, this broad critique of rationality—indeed, that it "will not save us"—not only comes from McNamara and the supplemental textual materials but also grows discursively from the image track of the film itself.

"LESSON 7: BELIEF AND SEEING ARE BOTH OFTEN WRONG"

Shortly after recounting his invitation to Washington by John Kennedy to serve as Secretary of Defense, McNamara (or the film; we're never sure which is structuring the chronological narration of the events) turns to his account of the Gulf of Tonkin incident, the discussion of which makes up the core of the second key lesson in the film: "Lesson 7: Belief and Seeing Are Both Often Wrong." Here, the film thematically and formally points back to the two segments analyzed previously by including the same archival shots of soldiers on a ship preparing for battle that accompanied the opening credits.

As McNamara recounts the miscommunication that led to the misperception that Vietnam had attacked the USS *Maddox* in the Gulf of Tonkin, the film cuts to an audio recording of a conversation between the Admiral U.S. Grant Sharp Jr. and General David Burchinal, who determine that the error was the result of a "mistaken sonar reading." Here, an archival image appears of three sonar men staring into a screen. As the error is revealed, the film cuts back to McNamara briefly as he recounts the chain of events that led from this event to the escalation of the war, which Morris pairs with original footage of a chain of dominoes falling across a map of Southeast Asia. McNamara reflects on the experience:

McNamara: It was just confusion, and events afterwards showed that our judgment that we'd been attacked that day was wrong. It didn't happen. And the judgment that we'd been attacked on August 2nd was right. We had been, although that was disputed at the time.

Ultimately President Johnson authorized bombing in response to what he thought had been the second attack—it hadn't occurred but that's irrelevant to the point I'm making here. He authorized the attack on the assumption it had occurred and his belief that it was a conscious decision on the part of the North Vietnamese political and military leaders to escalate the conflict and an indication that they would not stop short of winning.

We were wrong, but we had in our minds a mind-set that led to that action. And it carried such heavy costs. We see incorrectly or we see only half the story at times.

Morris: [off-screen]. We see what we want to believe.

McNamara: You're absolutely right. And belief and seeing, they're both often wrong.[21]

Here we have a chorus of voices: the voices of the two men on the phone, the voice of McNamara, the voice of Morris, and, of course, the visual "voice" of the images we see. The film rhetorically pairs the image of the sonar men staring into the screen with the image of McNamara staring into the camera, implicitly connecting their faulty observations with his own subjective point of view. The slow-motion shot of dominoes falling both alludes to the "domino theory" behind the escalation of the war and provides a visual metaphor of historical causality.[22] This connection between an ideological framework and series of errors suggests a causal chain between faulty observations and the unintended consequences that result from acting on such observations. Thus, while giving McNamara the final "word" (at least in the spoken sense) on one of the most debated events in the Vietnam War, the film simultaneously demonstrates that any individual interpretation is open to flaw and failure—a point that undercuts not just McNamara's perspective but also our own.

If the prior lesson demonstrated that "rationality will not save us," then its combination with these thoughts on "belief and seeing" becomes all the more alarming. The film's skepticism toward rational decision-making is expanded here to include skepticism about human perception and its ability to gather the proper information in the first place. This indeed is the point that McNamara and Blight, as well as Blight and Lang in the book, want to make about the event. As the text states: "How ironical and tragic—how absolutely surreal—that the August 4, 1964 watershed leading to a war in which three million people were killed was the result of a double misunderstanding."[23] The double misunderstanding in this case refers to both the error of the "overeager sonar men" and the misperception by the leaders of the two countries that this event and its response indicated a shared commitment to go to full-scale war.[24]

But as with "Lesson #2," the film expands the scope of the critique that McNamara offers to include its own larger claims. Rather than provoke a broad suspicion of

FIGURE 2.2. "Overeager" sonar men.

FIGURE 2.3. The domino theory in action.

observation, the film scrutinizes the specific forms of mediated "seeing" that we engage in via media technology. The segment therefore includes archival material of not just the sonar men but also the subsequent footage of Johnson announcing the attack on television and committing the nation to a justified response. Misperception thus occurs not only at the level of the individual and their given ideological mindset but also in the tools and technologies that we use to extend, record and transmit these perceptions to others.

This same theme reappears as McNamara narrates the events of the Cuban Missile Crisis in another of the crucial reenactments that structure the film. As McNamara discusses the inability of the US to determine definitively the presence or absence of Soviet missiles inside of Cuba, the film cuts to footage of large photographic transparencies of the aerial surveillance photos from the Cuban missile crisis illuminated by a series of light boxes. As the camera inspects the photographs alongside an unseen human observer, various lenses and magnifying devices pass over and in front them, distorting and manipulating their contents. At one point, we see an image of a human eye peering through a photographic loupe, the magnification from which gives the eye a bulbous, distorted appearance. Considered alongside the discussion of the Gulf of Tonkin incident, this sequence reveals that the view from above can be just as faulty as the view from the ground. Placed in the context of McNamara's revelation that the Kennedy administration had wrongly assessed the presence of missiles in the photos, the images illustrate that looking closer does not always mean seeing more clearly.

The reflexive nature of their content gives these moments in the film a special charge. As the film invites us via the cinematic apparatus to explore the perspective of a man who played a key role in history, it foregrounds the subjective nature

FIGURE 2.4. The observed and the observer in *The Fog of War*.

of human vision and questions the reliability of the technology we rely on to help us extend and improve this vision. While these sequences demonstrate that intelligence gathering in a hostile environment rests precariously on the limits of technology and the distortion of framing ideology, I would argue that the film extends this critique even further. That is, McNamara's reflections on the failure of "belief and seeing" in hostile environments give way to a larger critique of the relationship between reality and its media representations—a point I'll turn to now in considering the film's digital manipulation of its archival materials.

ANIMATING THE ARCHIVE

As the above demonstrates, one of *The Fog of War*'s primary concerns is the formal nature of the media we use to transmit information—often the same forms of media that Morris relies upon for much of the core visual material in the film. However, as I have also argued, these materials are not included simply to "illustrate" the content of McNamara's narration. Instead, these archival images form a visual voice that challenges, amplifies, and expands upon the claims of its subject. But their formal presentation also undermines their own claims. Even as Morris relies heavily on the archive to create the film, these historical records often communicate meanings that are decidedly different from those they originally expressed.

Consider, by way of contrast, the work of another documentary filmmaker credited with "bringing history to life" in his films: Ken Burns. Similarly laden with archival material, Burns's films earnestly attempt to collect and coordinate a wealth of historical material by pairing elements that will expand upon and reinforce one another. Archival photographs, panned and scanned in what has famously become known as the "Ken Burns effect," are accompanied by period music and the narration of letters, diaries, speeches, and newspaper articles from the time. This archival unity implies that a variety of media perspectives provide a sufficient representation of the past to comprehend its enormity.

In *The Fog of War*, however, the archival representations from the past are revealed to be not only fallible but fallible to a degree that undermines the evidence they provided in the past as well as the present. Thus, Morris's inclusion of this archival material seeks critically to unpack its pretensions and misperceptions to discover the sort of hidden truths that may lie beneath. This skepticism regarding access to the past is, of course, the thrust of the film. As McNamara states at the outset, "In my life I've made mistakes, but my rule has always been to try and learn, and pass these lessons on to the future." The film's contribution to this project is to question not just past events themselves but also the material residue they leave in their wake.

In part, this aim is achieved through the sort of selection and recontextualization that Jayne Loader, Kevin Rafferty, and Pierce Rafferty mastered so artfully in films like *The Atomic Café* (1982). *The Fog of War* similarly takes footage from any number of sources and recontextualizes it to illustrate the film's larger points. Whatever its original purpose, it seems unlikely that the footage of battle preparation from the opening credits was ever intended to question the ability of the military to gather proper intelligence, as I've suggested here. The film's inclusion of the outtakes from the press conference that open the film suggests a similar, subversive rereading of the footage's original intended meaning. In this sense, the industrial and propaganda materials that form the backdrop for the film all play unwitting roles in testifying to their own limitations and reveal their latent potentiality for remediation and reinterpretation. Such a move marks the film's unique utilization of the archive and sets its approach off from the earnest, good-faith quotation of a Ken Burns film.

At other points, however, the film goes beyond simply recontextualizing its source material to overtly manipulating it. Again, Bruce Conner, Craig Baldwin, and others have long utilized and manipulated archival material to critique and undercut its original rhetorical use. But unlike other found-footage films, *The Fog of War* combines this material with the testimony of an eyewitness observer. Consider, for example, Bruce Conner's use of found footage and media coverage in *Report* (1963–67). Conner's juxtaposition of the footage from John F. Kennedy's funeral procession with battle footage and a bullfight offers a startling, subtle critique of a society that thrives on the media-driven spectacle of violence.[25] While Morris's work clearly shares political sympathies and formal methodologies with Conner's biting, ironic media satire, he differs from Conner in his utilization and juxtaposition of this archival material with the first-person interviews of his subjects. *The Fog of War* thus seeks a middle ground between the earnest archival

FIGURE 2.5. The "number cruncher" becomes the bomber.

unity of Ken Burns's work and the ironic self-reflexivity of Conner's assemblages. Neither entirely redemptive nor dismissive of the archive, Morris takes a unique approach to these materials by digitally altering them at key moments to punctuate and critique McNamara's thoughts. Digital alteration—usually associated with undermining the truth or faking it—here suggests that such transformations can reveal the truth.

In what has become one of the film's more notorious segments, McNamara relates how he and General Curtis LeMay arrived at the means and methods for firebombing Japan. As Morris has claimed, this is the first place where McNamara discussed his participation in these events—events that many consider to be tantamount to the eventual choice to drop the atomic bombs on Hiroshima and Nagasaki. After stating that an operation had burned to death "one hundred thousand civilians—men, women, and children—in a single night," Morris asks McNamara if he knew this was going to happen. He replies, "In a sense, I was part of a mechanism that recommended it." At this moment, after having chaotically flipped through documents, photographs, and images from the period that document the missions, the film cuts to an image or footage (we aren't sure what's causing the movement) of animated blue numbers and statistics falling out the bomb-bay doors of an aircraft down onto a city below. The original source material, a black-and-white, sepia-toned photo, is identical to countless others that feature bombs falling out of an airplane, but this one overtly implies that the use of statistical rationality was equally damaging. Again, whatever its original purpose, through the use of CG animation the photograph becomes the film's most direct indictment of its subject.

Shortly after the "falling statistics," McNamara describes a report he wrote for LeMay that argued for flying the B-29s at a lower altitude during their bombing missions. While this decision increased the risk of a plane being shot down, it dramatically increased its effectiveness in "target destruction." Utilizing a technique that Morris has described as "3-D photography,"[26] the film cuts to a black-and-white image of bombs dropping from a plane. The camera appears to zoom in to the image, but rather than simply enlarge the elements equally as a typical zoom would, elements in the foreground appear to expand and move more rapidly out of the frame relative to those in the background. The visual effect not only yields the impression of three-dimensionality that Morris describes but also gives us the feeling of dropping out of the plane alongside the bombs themselves. In a sense, this is exactly what the men who piloted the planes were doing, given that, under McNamara's direction, they lowered their flight altitude to the extent that they became targets themselves for Japanese antiaircraft fire. While the shot lasts only approximately eight seconds on-screen, the 3-D effect is startling enough to call it out among the dozens of similar images that the film contains and marks the significance of this portion of McNamara's testimony. As the image digitally "comes to life" relative to the others, we gain the sense that McNamara has gone from being a witness of history to one of its actors, directing its outcome rather than passively observing its course.

One final instance of digital manipulation further illustrates Morris's approach to his archival material. Although less technically innovative than the previous two, its effect is no less powerful. This moment comes as McNamara discusses the result of the firebombing that LeMay carried out on Tokyo and the devastating impact the bombs had on what he calls "a wooden city." Morris's voice is heard off-screen asking McNamara: "The choice of incendiary bombs, where did that come from?" McNamara replies to the effect that the problem lay not in the method of destruction so much as in its extent. He goes on to list the other cities that were similarly destroyed, comparing each target to a similarly sized American city. He states: "[LeMay] went on from Tokyo to firebomb other cities: 58 percent of Yokohama, Yokohama's roughly the size of Cleveland; 58 percent of Cleveland destroyed . . . 99 percent of Chattanooga destroyed, which was Toyama; 41 percent of the equivalent of Los Angeles, which was Nagoya." As he lists the cities destroyed, a black-and-white photograph of ruins appears, with the name of the Japanese city and the percentage destroyed superimposed in red text. This black text fades, giving way to the name of the US city in black text over the same photo. At first, the technique simply illustrates McNamara's examples, but once he stops with the list above, the image track goes on, listing dozens of other cities at an accelerating pace in time with the music. McNamara's point is certainly powerful enough on its own. But combined with the effect of the extended list and its chaotic, accelerated pace, it becomes ample evidence of McNamara's admission, at the end of the sequence, "that [Lemay], and I believe I, were behaving as war criminals."

Graphic superimpositions of this sort are nothing new, but their use here nonetheless stands out for the ambivalent position they occupy between McNamara's

FIGURE 2.6. Tokyo 51.0%. Graphic superimpositions of the percentage of devastation for each city firebombed in World War II provide a powerful combination of two information sources: data and photography.

message about the past and the film's message about him. They pose a contrast between two forms of evidence and representation: the statistical and the photographic. The images of devastation are sufficiently generic that they simply become signifiers of the concept itself rather than descriptors of a given event. Their historical specificity and emotional connection to the audience derive entirely from the names and numbers affixed to them. And yet, the film simultaneously calls this type of statistical information into question, or at least aligns it with the rational worldview that brought about this devastation in the first place. Thus, the statistical information also lacks a level of historical sufficiency without a view toward the physical devastation that it corresponds to—a dimension provided by the images that form the backdrop.

This series of images and their superimposed identifiers occupy a curious middle space. On one hand, the statistics represent the startlingly calculated rational efficiency with which the destruction of Japan was carried out. (As McNamara states, LeMay was the only general who focused exclusively on the percentage of target destroyed per unit lost.) And yet, the film pairs them with photographic representations to redeem and represent that loss by powerfully conveying its true extent. What was once used for the rationalized optimization of destruction (statistical quantification) is now used to generate commemoration and empathy. That which had faded into generic, historical obscurity (photographic evidence of the devastation) is once again rooted into historical time and space. At the time of their creation, such representations were utilized to document and perfect the destruction that they quantify and capture. In retrospect, these same representations stand as evidence of the guilt of both McNamara and LeMay by documenting their crimes and reinforcing the extent of their impact. While neither form of representation—statistical quantification or photographic evidence—is sufficient on its own to reach this conclusion, figured together in this series of superimpositions they reinterpret one another and provoke a self-consciously synthetic visualization of this untold moment in the history of the war.

THEORIES OF HISTORY AND THE ARCHIVE

Along with the falling statistics and the 3-D animations, these graphic superimpositions demonstrate the film's ambiguous approach to its archival material and interview subject. I say "ambiguous" because although there is a reliance on the archive to represent the past, its constant manipulation throughout the film betrays a clear suspicion about its ability to self-sufficiently convey historical truth. Of course, even terms like "historical truth" and "representation" are notoriously slippery and ambiguous, opening themselves to extensive debate by credentialed historians and theorists about the existence of objective truth and its ability to be captured or represented in any given form of history.[27] Despite this ambivalence, however, the film nonetheless approaches its subject with a definite theory of truth and history. Academics may not have agreed on the existence of an objective reality or the possibility for unmediated, individual access to it, but Morris as a filmmaker clearly believes in both propositions. For example, in the June 2000 interview with *Cineaste* cited earlier in which he discusses the tendency of people to "live in a cocoon of one's own devising,"[28] he contrasts this tendency toward individual, subjective delusion with a resolute belief in objective reality. Responding to a question about his background in philosophy and the influence of thinkers like Foucault on his work, his response is worth quoting at length:

Morris: I'm certainly aware of it. But my background is in American analytic philosophy rather than in Continental philosophy, and that's where my sympathies lie. I once said that one of the good things about Cambridge, Massachusetts, is that Baudrillard isn't in the phone book. Because first and foremost there is a kind of realism behind all of the movies that I've made. Realism in the philosophical sense. That there is a real world out there in which things happen. . . . This is not up for grabs. You don't take an audience survey.

Cineaste: So we have an unmediated relationship with the fact.

Morris: I wouldn't say that our relationship with the fact is unmediated, but there is a fact out there.

Cineaste: But we have direct access to it.

Morris: Well, the world leaves a trail, and it is our job as investigators—or, specifically my job as an investigator—to try to lead myself back to the world. It's not something that you just grab hold of. . . .

[W]e know about the world, we know about our history, through the things that history has cast off, whether it's pieces of evidence, documents, the testimony of people who have lived through those times. . . . History comes by only once, and the residue of history can be lost.[29]

Although using interviews to interpret a film risks confusing textual meaning with authorial intention, this is an instance where such statements merit a little scrutiny. As a "conversation" between McNamara's words and Morris's images, the film explicitly addresses competing theories of history, and its release alongside a book and countless other interviews testifies to a desire to make this theory explicitly part of the film's reception. While Morris's theory of history may not be identical to the film's, it at least forms part of its backdrop, and this exchange clearly demonstrates the interplay between the archive and McNamara's testimony that I have been describing in the film.

The theory of history that Morris puts forward offers historical truth as a possibility, but a fragile and fleeting possibility that must be delicately unearthed through diligent investigative efforts. On one hand, Morris claims, individual social actors have the potential to delude themselves about "reality" and construct for themselves "a cocoon of their own devising"—a possibility shared by McNamara and Morris in the contention that "belief and seeing are both often wrong," and one more than amply demonstrated in Morris's prior films *The Thin Blue Line* and *Mr. Death*. And yet, he also argues for the potential of critical reflection by an eyewitness to provide one of the "pieces of evidence" that make up the "residue of history."

But such testimony is only one piece of the puzzle. Hence the need for the other forms of evidence that history has "cast off," from documents and photographs to archival footage and statistics. This archival focus on different types of media is what partially differentiates both *The Fog of War* and *Standard Operating Procedure* from Morris's earlier work (although there are similarities as well). At the same time, however, none of these individual records—the "trail" that history has left behind—sufficiently leads us "back to the world." This lack necessitates their critical evaluation and reassemblage in the film. Taken as a whole, Morris believes these revised sources may lead back to some level of historical truth, although even when they are preserved and present, the truth they offer is far from self-evident. Individual testimony, historical documents, and archival materials on their own are insufficient. But when reworked, digitally manipulated, critically interrogated, and contradicted, these materials contain a latent potential for representing the past.

A level of skepticism regarding the self-sufficient transparency of the past operates in the background of the film and its treatment of the archive. The film's dense collage of archival material is animated (or reanimated) in a manner that interrogates its specific historical truth but also the archival impulse more broadly. One gathers from the film's eagerness to tinker with these materials that the "residue" of the past collected in the archive is perhaps a necessary condition for achieving historical truth, but not a sufficient condition in its own right. The investigator—the one who will seek out and critically interrogate the evidence—is also essential to the process. If we are to achieve historical truth via the archive, if we are to lead ourselves "back to the world," we must tease out this truth from a mass of material in which truth is anything but self-evident.[30] "It's not something you just grab hold of," as Morris puts it.[31]

Interestingly, however, the film arrived at a moment when our culture was witnessing an extreme bout of "archive fever," to borrow Jacques Derrida's phrase.[32] One of the fastest-growing portions of the Internet before the rise of user-generated content on social media was the digitization of existing analog archives. This effort was motivated by the hope that putting these materials online might finally arrest the process of physical decay, thereby transforming them into durable, universally accessible resources.[33] But, as Wendy Chun points out, the digital technologies that the computer and the Internet comprise were, from their very conception, viewed as tools capable of organizing the world's information, long before Google took this as its mission.[34] As early as Bush's "As We May Think" and John von Neumann's "First Draft of a Report on the EVDAC," a desire existed for a living, accessible archive of information.[35] But the various technologies we use to achieve this goal are universally reliant on regenerative repetition—a quality that makes them more similar to human memory than archival storage. Rather than a permanent, accessible archive of all the world's information, the experience of the online archive is one of broken links and missing files on a micro level and the medium-specific churn of old and new material on a macro level. The archival Internet is at once a place of both memory and forgetting, creation and deletion, a state Chun calls the "enduring ephemeral."[36] "New" material seems instantaneously outdated, and old material is constantly rediscovered and recirculated as new.

Morris's concerns about the "perishability" of history in his *Cineaste* interview, as well as *The Fog of War*'s general thrust to draw lessons from the past, both align with the positivist, archival thrust that Chun locates in the drive to digitize. And yet the film's critique of McNamara as an "IBM machine with legs" and its willingness to digitally tinker with rather than faithfully transcode its archival sources point to a certain skepticism regarding the transparency and self-sufficient utility of the archive, digital or otherwise. Of course, the film is not "about" the digital archive but rather its critique of computer-driven logic and rationality. Its contention that "rationality will not save us" points to an awareness that there is more to unearthing the past and unlocking the truth in the archive than simply transcoding it into a digital form.

Perhaps the best way to characterize the historical theories of Morris and the treatment of history in the film is by way of reference to an existing if not mainstream approach to visual historiography advocated by the art historian Stephen

Bann. Bann's work charts the rise of what he calls "historical consciousness" in the visual culture of nineteenth-century Europe. Drawing from Hayden White's tropological theory of historiography, Bann contends that this growing historical consciousness over the last two centuries has delivered us into an era of post-modern irony regarding the visual presentation of history in venues ranging from Colonial Williamsburg (which effaces the difference between present and past in a move not unlike a Ken Burns film) to more overt, self-conscious juxtapositions of multiple temporalities like the work of landscape architect Bernard Lassus. Lassus's work restoring historical spaces seeks to preserve the present of the space together with, in Bann's words, "yesterday, and the day before yesterday"[37] in such a way that all are simultaneously present and yet faithful to the individual periods. While such juxtapositions might seem confusing, Bann argues that contemporary spectators have developed the faculty of "seeing double"—that is, holding in their vision multiple sites of historical engagement at once.[38]

The notion of an "ironic museum" in which past and present are preserved in their temporal and formal separation but sit self-consciously and playfully side by side perfectly captures the approach to the archive that we see in *The Fog of War*. This ironic gesture of juxtaposition without reconciliation helps square the film's various paradoxical positions: a critique of computational rationality presented using extensive digital effects, camerawork that reveals the biased nature of the camera itself, a man reminiscing about the fallibility of human memory and perception. Like a museum that places today and yesterday (and the day before) side by side, these points are laid out but not reconciled. The film's insistence that the past is worth preserving and contains lessons for the present saves its ironic methodology from devolving into parody or pastiche. Even as the film's opening footage reminds us that all media are manipulated, there is a gravity to its tone and subject matter that compels our attention. Indeed, the film's manipulation of its source material continually reminds us that the "truth" of images is never entirely immanent to the media themselves; rather, truth derives from the rhetorical and critical contexts in which media appear. At once distrustful of the archive but reliant upon it, dismissive of logical rationality but earnest in pursuing some level of historical truth, the film sits evenly between an abstract meditation on media and an exploration of the past that it has captured—a divided attention that will carry over to Morris's next cinematic project, *Standard Operating Procedure*.

STANDARD OPERATING PROCEDURE'S
IMAGE AESTHETICS

If *The Fog of War* works as a meditation on the archive writ large that draws on many forms of media from many different sources, then the focus of the archive in *Standard Operating Procedure* is far more closely circumscribed. Rather than exploring charts, graphs, reconnaissance photos, news footage, audiotapes, newspapers, and other media as *The Fog of War* does, *Standard Operating Procedure* turns its attention to one specific form of media—the digital photograph—as it is instantiated in one specific collection: the images that emerged from the Abu Ghraib prison complex in Iraq in April 2004. In spite of this shift in scale, however, *Standard Operating Procedure* continues *The Fog of War's* exploration of the collision between historical events, social actors, and the media representations they leave behind. Like *The Fog of War*, the film is as much about the media representations of an event as it is about the event itself.

Indeed, the two films share a sort of inverse, mirror relationship with one another in several other ways as well. Morris himself calls *Standard Operating Procedure* the "flip side" of *The Fog of War*, "because instead of a policy-maker—perhaps the most important person in the government save the president himself—here you have grunts, people with little or no power,"[39] a point Linda Williams echoes in her discussion of the film.[40] Beyond the difference in rank of their subjects, both films are obviously about war and its effect on both perpetrators and victims, and both films explore the media that these conflicts produce. Furthermore, while *The Fog of War* was a critical and commercial success, receiving generally positive reviews in the mainstream media and garnering Morris an Academy Award for Best Documentary Feature, *Standard Operating Procedure* received mostly negative reviews from critics and went on to fail miserably at the box office.[41] But if *The Fog of War* received more attention than *Standard Operating Procedure* in the mainstream popular press, in academic circles the situation was reversed. Since its release, *The Fog of War* has been largely ignored in journals and other publications, whereas *Standard Operating Procedure* has generated a great deal of controversy and attention from film and media scholars at conferences and in publications.[42]

Beyond their reception, the two films also mirror one another in that both were released with an eponymous companion text. But whereas Blight and Lang's text expanded the historical facts and philosophical issues explored in *The Fog of War*, Philip Gourevitch's text instead offers a narrative account of events leading up to the Abu Ghraib scandal and its aftermath.[43] For their source material, Blight and Lang drew from their own preexisting research (generated over a decade of working with McNamara before he sat down with Morris). Gourevitch, on the other hand, derived his text largely from the material Morris himself collected for the film. (The interviews alone ran to almost 2.5 million words.[44])

The treatment of both films on Morris's website also offer parallels and differences. Morris, in the four years between his two films, was slowly becoming a multimedia artist, using the web to *expand* his films rather than just promote them. Whereas errolmorris.com largely followed *The Fog of War's* release as it was covered by other media (collecting reviews, release events, and interviews with Morris), for *Standard Operating Procedure* the site began to feature original content that explored points in the film further and defended Morris's actions in several of the controversies that erupted during its theatrical release. While much of this new material was also part of the blog Morris began writing for the *New York Times* in September 2007, other material on the site related to the film is unique to the site itself (e.g., the sections "The Grump" and Morris's thoughts on several of his "Aborted Projects"). Furthermore, as *Standard Operating Procedure* comes to focus on an exclusively digital medium—photography—his own "digital" activities online begin to expand as well.[45]

The shift from *The Fog of War* to *Standard Operating Procedure* is thus not a clear thematic break, but rather a shift in focus and scope. Instead of focusing on the life of a single individual who had a hand in several of the bloodiest and most technologically mediated wars of the twentieth century, *Standard Operating Procedure* meditates on the role of a specific media technology in relation to a specific event. But if Morris tightens the focus of *Standard Operating Procedure* to a single technology and event, the problem he explores—namely, the role of photography in our understanding of an event—is approached on a number of fronts at once. In addition to the film, Morris begins simultaneously exploring these ideas on his blog, and considering his work there alongside the film expands our understanding of both.

THE OPINIONATOR: MANY THOUSANDS OF WORDS

Tellingly, Morris's first post to the *New York Times* blog *The Opinionator* appeared nearly a year before *Standard Operating Procedure* premiered, but its content clearly reflected what must have been a major preoccupation at the time given the film that he was in the midst of making. Entitled "Liar, Liar, Pants on Fire," it offers a discussion of the possibility for photographs to be faked and the role that context plays in their reception and interpretation—an issue that would return front and center once the film came out. It begins: "Pictures are supposed to be worth a thousand words. But a picture unaccompanied by words may not mean anything at all. Do pictures provide evidence? And if so, evidence of what? And, of course, the underlying question: do they tell the truth?"[46] This post offers a fitting preamble to the blog itself, and subsequent posts deal further with photography and reenactment, perception, memory, and any number of other issues central to the investigation in *Standard Operating Procedure*. Many of the posts are extremely long by blog or even newspaper-article standards, often running to thousands of words and spread out over several installments. In one post, Morris even acknowledges, in response to reader comments, that he's not blogging so much as posting essays—a point he admits before comparing his own method to Descartes's in the latter's *Meditations on First Philosophy*.[47] This, moreover, stands in marked contrast to the laconic presence within his films, where he speaks little, if at all, and offers no narration or voice-over. If, as his first post reiterates, a picture is worth a thousand words, then he seemed focused on using the blog to give the photos he discusses their textual due. Given its thematic preoccupations and its simultaneity with the production and release of *Standard Operating Procedure,* the blog thus forms an additional if indirect background text to the film.

Unlike the focus of the film, Morris's thoughts on photography in his blog only occasionally turn to the Abu Ghraib photographs. Morris's subject is more generally the issue of truth and photographic representations—an issue that leads him to explore the work of Roger Fenton, Matthew Brady, Walker Evans, and others. In typical Morris style, his posts generally begin with a series of older archival images. He then poses a series of questions the images raise upon closer inspection in the tone of an investigation or a detective mystery. These questions often relate to the historical circumstances surrounding the photos and the extent to which they can be said to reflect the "truth" of the scenes they capture. Morris the blogger and Morris the filmmaker draw on a similar set of ingredients: equal parts quirky detective fiction and meditative philosophical reflection on the nature of reality/representation and history/memory. Given the nature of his films, it is not surprising that he often gravitates toward the eccentric and bizarre sides of subjects. A post on anosognosia (the lack of awareness about one's own illness or impairment), for example, begins with an anecdote about a bank robber who covered his face in lemon juice, mistakenly thinking this would allow him to remain invisible to the security cameras that were eventually used to apprehend him.[48] But beyond mirroring the style of his films in general, the material on his blog often relates directly to the issues addressed by *Standard Operating Procedure.*

In one of Morris's first posts, for example, he takes the two Roger Fenton images from the Crimean War entitled "Valley of the Shadow of Death" that have been discussed by Susan Sontag and others and proposes that one of the two nearly identical images must have been staged.[49] Calling them "ON" and "OFF" in reference to the placement of a series of cannonballs in the middle road, Morris investigates a number of different techniques to determine whether Fenton or another party moved the cannonballs into the road or into the ditch for the second image. As a choice of topic, the Crimean War is a natural one in that, as Ulrich Keller has noted, it represents a sort of transitional stage in the visual history of warfare.[50] On one hand, it was the last war to be fought as a grand spectacle for the eyewitness observer, since modern weapons like the machine gun made bold charges toward the enemy dangerously obsolete. But on the other hand, it was the first war to be thoroughly visually documented by modern forms of media representation like lithography and photography. Fenton's photographs, then, are the first to be taken of any war ever, and Fenton, as Sontag notes, is repeatedly cited as the first war photographer. None of this is lost on Morris, as he cites Sontag's book repeatedly, and interviews Keller himself. The blog thus reveals the degree to which Morris researched the relationship between war and photography as he prepared to make a film about the Abu Ghraib scandal.

Beyond their status as historical forerunners to the Abu Ghraib photos, the Fenton photographs are also relevant to the making of *Standard Operating Procedure* given the nature of the questions they pose. That is, once we allow that the scene on the hillside was altered for one of the images, we must immediately ask which image and why. Leaving aside most of the intricacies involved in Morris's attempt to order the images temporally (suffice it to say it takes him nearly nine thousand words and the use of spectral analysis to do so), it is worth noting that he traveled back to the location where the images were taken to record his own images and reenact the conditions of their capture—an effort not uncommon in his film projects, and one that led to a great deal of the criticism of *Standard Operating Procedure*.[51] Moreover, the images provide Morris with an occasion to reflect on Fenton's motivations for altering the landscape of his subject. Was he trying to put the cannonballs back in the position where they would have originally landed? (In other words, was Fenton himself reenacting the scene?) Or was he simply trying to capture a more dramatic shot? (And would that have consequently been more or less faithful to the subject he was attempting to capture?) In essence, Morris is concerned with the interplay of visual aesthetics and factual reportage in the two images and which version was more faithful to the veracity of the subject Fenton felt it was his charge to document.

With both the Abu Ghraib photos and the Fenton photos, Morris delves into images of war that were staged or acted out for the benefit of the camera. As many commentators have noted, there is a complicated co-incidence in the Abu Ghraib images between the presence of the camera and the acts of torture that it records.[52] On one hand, the absence of a camera would deprive the world of evidence of these acts, so the camera and its images are necessary to understand what took place. On the other hand, there is a great deal of evidence that some of the forms of torture documented by the camera were specifically staged to create a visual spectacle for the benefit of the camera itself. Thus, what happened *before* the camera might not have happened *without* the camera (or at least not in the same fashion). Paraphrasing Morris's title from the Fenton post, we might ask, "Which came first, the spectacle or the camera?" But like Fenton, the perpetrators of the Abu Ghraib images arranged the scene in a certain fashion for maximum dramatic impact. Summing up his search, Morris takes a moment to wax philosophic about his desire to arrange the images:

> I sometimes wonder: is the entire meaning of photography contained in these twin Fenton photographs—one the *doppelganger* of the other and often indirectly described as such? The good Fenton photograph, honest and unadorned by a desire for contrivance or misdirection, and the bad Fenton photograph—the photograph decried by Sontag—corrupted by the sleight of hand, the trick, the calculated deception.
> But which is which?[53]

In a sense, the Abu Ghraib images present a quandary because they occupy the space between the two Fenton images, and that perhaps is why they came to occupy Morris in the first place.

I raise the issues presented in the blog because how we read that content in relation to *Standard Operating Procedure* affects how we interpret the aim of the film, and, as I will argue, this is a film in which context and the classification or categorization of an object is very much at stake. That is, if we see *Standard Operating Procedure* as an investigation into the Abu Ghraib prison scandal and the question of US policy on torture, then we are inclined to place it alongside other films dealing with similar issues, like Alex Gibney's *Taxi to the Dark Side* (2007), Michael Winterbottom's *The Road to Guantánamo* (2006), and Rory Kennedy's *Ghosts of Abu Ghraib* (2007).[54] This, of course, is perfectly appropriate given the subject matter and thrust of the companion text and the simple fact that *it is a film* by a well-known filmmaker. But if we place the film in the context of Morris's previous film on war, *The Fog of War,* and his other activities on his blog and elsewhere, then the subject matter takes on a different valence entirely. Seen as part of an ongoing meditation on the relationship between representation and reality, photography and the external world, the film is less about specific policies and events or individual culpability and more about the nature of perception, representation, and human behavior. As the content on the blog indicates, the role of photography in warfare and the nature of photographic technology in documenting and interpreting such momentous events are topics that occupy Morris far beyond any one particular instance or set of photos. We might conclude, then, that *Standard Operating Procedure* is not so much about Abu Ghraib the historical event as it is about the Abu Ghraib images, and their role in the event.

How one determines the film's true focus seems to dictate the extent to which one finds any merit in the film's overall project or approach. Returning to the controversy the film generated, we can draw a fairly clear line between those who did or did not "like" the film based on what they thought its overall subject and intentions were.[55] Scholars who fall into the latter category, like Bill Nichols and Irina Leimbacher, for example, read the film as being about torture and the circumstances behind the events captured in these images.[56] Given this, they find Morris's treatment of the images and his method of reenacting the torture sequences they depict to be fraught with a fetishized aestheticization of the events that lacks a moral center. Such critics further assert that his signature Interrotron interviews simply provide an opportunity for the perpetrators to deny ultimate culpability. Nichols's three primary objections, which nicely sum up the general reaction against the film, are: (1) the limited perspective of the guards, and their inability to assume any of the guilt; (2) the aestheticized nature of the reenactments; and (3) the absence of any voice for the victims.[57] His is a trenchant critique of the film, and if the film is about acts of torture, then all of Nichols's claims are indeed accurate and the film's flaws are, to some extent, inexcusable.

But if we shift the focus of the film from being about the event of Abu Ghraib to being about the images it generated, our reading of its method, and perhaps its faults and omissions, also shifts. Consider, for example, the description Julia Lesage (who was largely positive on the film) offers of its subject:

> I use a textual analysis of *Standard Operating Procedure*, which takes as its topic just the Abu Ghraib photographs, to explore issues of affect in the torture documentary. However, I also explore how the film works as an analytic documentary, one that explores what the photograph, or indeed witnesses, can and cannot convey. *Standard Operating Procedure* particularly raises the question of "authenticity" in relation to its interviewees. We are asked to evaluate not only the history of Abu Ghraib torture that these participants tell us about but also how much we trust what they have to say.[58]

Lesage clearly feels that the film is about "just the Abu Ghraib photographs" and what they or their creators "can and cannot say." Linda Williams similarly reads the film as an interplay between the images and their creators, insisting that the images have as much to do with the larger ideological context that exists as they do with the frame they impose.[59] In a reading that lies closest to the one I am proposing here, Caetlin Benson-Allott writes:

> *Standard Operating Procedure* focuses on how atrocities become media files. Morris's film asserts that although the abuse at Abu Ghraib is undeniably terrible and true, the photographs neither speak directly to us nor offer transparent access to the events. The photographs are insufficient and require interpretation from viewers, who may bring external impressions and motivations to the task. *Standard Operating Procedure* tries to communicate this problem by focusing on how mediation, and digital mediation in particular, disorients rather than facilitates our processes of interpretation.[60]

Like Williams and Lesage, Benson-Allot determines that the film focuses on the subject of digital photography and the photographs themselves and that, in this particular arena, the film offers an important, worthwhile intervention and addition to the collection of films on the Iraq war.

Interestingly, none of the scholars who praise the film deal very extensively with the reenactments it contains. Their discussion of the interview segments (which, for Nichols, allowed the subjects to deny guilt) emphasizes the way in which the film forgoes the question of guilt, leaving this for the audience to decide. My aim here is not to determine which side of the debate is correct or incorrect (though the thrust of the reading I'm offering obviously aligns more closely with those who think the film is about photography and mediation). While I agree with Benson-Allot and the others that the film is ultimately about this collection of digital images, I believe that the emphasis here is not on the images per se but on the collection itself. That is, the extensive commentary thus far offered on the film largely misses the images' status as a *database* of images.

DATABASE AESTHETICS

Almost without fail, nearly every critic of the film points out that the images that it deals with are "digital" rather than analog photographs, and that this fact has something to do with the mutability and transportability of their contents. Had they been analog images, they would have been far easier to contain and perhaps less likely to have been created in the first place. Digital images are at once more and less private. Lacking the need for a third party to develop and print the negatives, they can reveal their contents but remain the exclusive property of their creators. And yet, digitality also facilitates copying and sharing, lending them an instantaneous ubiquity that analog photos lack. Digital photos, moreover, are far more open to manipulation via programs like Adobe Photoshop. All of this is surely accurate, and as we will see, the film does highlight their status as *digital* images. And yet, the film doesn't emphasize either of these particular properties, even though both lend themselves to the sort of questions Morris often addresses. Rather, *Standard Operating Procedure* emphasizes and questions another facet of their role as digital media: their status as a collection of files, or, more accurately, as a database. Interrogating the database, Morris most clearly advances the larger themes of representation, mediation, and truth that became so evident in his blog posts.

As it was with *The Fog of War*, the opening sequence of the film is telling. As the opening credit sequence rolls, or rather floats, the viewer is immersed in a cloud of spatially diffuse images floating back and away, a double movement that yields the impression that, as we drift steadily forward, our attention is directed stubbornly backward at images fading slowly into the distance. While many are immediately legible as the more iconic images from the Abu Ghraib scandal, they appear here robbed of any framing context but the frame itself. But what interests me here is not the images themselves or the frame around them, but rather the blank, nonrepresentational space in which they appear—a space that is rather overtly rendered as "no place." While focusing on this blank space instead of the sensational content of the images will at first seem counterintuitive and perhaps the epitome of disinterested spectatorship (how could one *not* look at them?), their distinct aesthetic milieu in the sequence foregrounds the film's relationship to the controversial material it explores.

The aesthetic of this dark nonplace can be illuminated with the area of digital art known as database aesthetics.[61] As far back as his influential *The Language of New Media,* Lev Manovich described the database as the dominant symbolic form of the digital age. It provides a new interface to the cultural field and replaces the centuries-long dominance of the narrative form that appeared in older media such as the novel and film.[62] While both forms, the database and the narrative, have always existed alongside each other, he argues that at different points either form rises to prominence—an exchange currently taking place thanks in part to the widespread adoption of the computer as the "Universal Media Machine."[63] The database as a cultural form is characterized as a collection of discrete entities with an infinite number of possible connections to each other but lacking in any necessary connections that order or prioritize these items. Unlike the narrative, which imposes specific cause/effect, beginning/middle/end relationships on its constituent elements, the database leaves these connections undefined.[64]

Returning to the opening credit sequence, the cloud of images that float there before the viewer present themselves, in this reading, as a collection of discrete digital records, which, as products of various digital imaging technologies, they undoubtedly were/are. Presenting them as a random cluster with no immediate logic to their arrangement or spatial distribution renders aesthetically the material status these records had at various points in their existence, from the nonlinear editing software that rendered any particular shot, to the hard drives of the computers to which the images were downloaded, to the memory cards of the cameras with which they were originally recorded. As digital files, they can be ordered according to any number of different principles—foregrounded or elided depending upon any number of preferences.

We can see this same aesthetic principle at work at several other points in the film. The discrete nature of the image as individual record, for instance, is foregrounded most explicitly in the discussion by army investigator Brent Pack about "metadata." He defines metadata as the "fancy two-dollar word for information about information" that allows him to order the images according to various factors including the date they were taken and the specific camera that captured them. The collection of images is, at other points, foregrounded as a database of such records. When Pack describes the beginnings of his investigation, he reveals that the army gave him twelve CDs' worth of images, which he then began to go through and organize. Here, the image track explicitly illustrates a screen with the "thousands of images from Abu Ghraib" as tiled icons on an apparently enormous screen. His goal, as he states it, was to find the images that depict prisoner abuse and to identify who might have been in the area at the time. As the screen rapidly flips through these records, sound effects reminiscent of a hard drive spinning click frantically away. As Pack focuses his attention, the screen isolates specific images, aesthetically calling them forth from the cloud; they appear as records pulled up from the database with individual labels enumerating the aforementioned metadata.[65] As he describes organizing the photographs according to various criteria, the screen image responds by arranging and rearranging images into various timelines.

Again, the material form of this collection of images outside the film *is* a database. This is not, therefore, a quality the film imposes on them. Instead it works to retain and foreground this materiality in their aesthetic treatment each time they reappear. The various visual and sound effects that connote the database here in

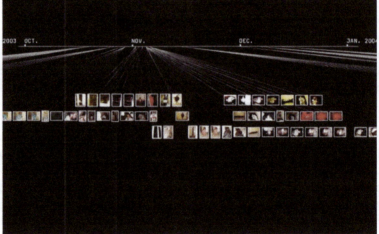

FIGURE 2.7. The database of images rendered in different aesthetic configurations.

this narrative medium are somewhat the reverse of the artificial shutter sound that plays when one snaps a picture on a digital camera: aesthetic, sensory appendages held over from another medium to remind one of their origins. Pack's investigation as it is presented in the film is this migration from one form, the database, into another, the linear narrative. As he puts it, "The pictures spoke a thousand words, but unless you know what day and time they were talking, you wouldn't know what the story was."

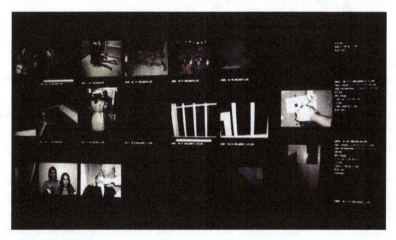

FIGURE 2.7 *continued.*

WHAT THE STORY WAS

Again and again the various social actors in the film highlight this same tension between the extreme legibility of what the images depict and their inability, as Pack says, to narrate the story adequately. This is the tension inherent in the database/narrative distinction that Manovich makes. As critics of New Media art point out, much of what constitutes database art in the strictest sense often presents itself as a "choose your own adventure"–style set of individual materials for participants to use in creating their own narratives. And this is largely what happened with the database of images from Abu Ghraib. Once they surfaced from the prison, any number of individuals and media outlets began selecting specific images and placing them into various discursive contexts, from the army investigation that Pack started in early 2004 to the *60 Minutes* broadcast that eventually introduced the scandal to the public. Indeed, the film itself is an attempt to understand "what the story was" that produced these images.

I would like to linger on this question of the insufficiency of the images themselves to stand for and represent the events they depict in a complete and self-evident fashion. This is, after all, the ineffable paradox of still photography: on one hand, so automatically, irrevocably indexically bound to the historical world, and on the other hand, so mediated, insufficient, and misleading about that world. Taken together, these fragments of time fail to offer a sufficient account of the circumstances of their creation. As a collection, they are, in the strictest sense of the term, nonrepresentational. While each individual image may offer a mediated, representational glimpse of what existed before the camera at a given moment in time, the photographs' collective meaning has to be supplied externally. The database allows us to order its contents according to any number of criteria, to declare certain images relevant and others irrelevant, and to classify them into categories like "criminal act" and "standard operating procedure," but these organizational schemas are necessarily external to the database itself. In database terms, the individual record itself may be representational, but the data set as a form is manifestly nonrepresentational. It can contain information, but meaning has to be found elsewhere. This is precisely what the film reveals in the extended CGI sequences that I describe as a form of database aesthetic.

Though the film foregrounds the plasticity of its source material, it does not, however, evacuate it of meaning entirely. Quite the opposite. By translating the database into a linear narrative, the film utilizes any number of techniques to account for the structure the database lacks, including the interviews, Sabrina Harman's letters, and, notoriously, the reenactments. Interestingly, the film lacks entirely those elements so omnipresent in *The Fog of War*. Rather than a broad collage of external media sources and archival documents, the film focuses exclusively on the Abu Ghraib images and supplements them with interviews, letters, log books, and reenactments. Instead of animating the contents of the archive by digitally manipulating their appearance or content as he did in *The Fog of War*, Morris aestheticizes the archive itself in order to foreground its immaterial, mutable nature.

The lack of an external, secondary media context of the sort that we saw in *The Fog of War* is, of course, supplied in our contemporary context by the flurry of media coverage that surrounded the Abu Ghraib images when they first appeared and were widely taken up and debated from any number of perspectives. For some, the Abu Ghraib images represented the work of "a few bad apples." For others, they offered a glimpse of the moral vacuity at the heart of the Bush administration's prosecution of the war on terror. For still others, they were the unsurprising proof of Western aggression against the Middle East, a manifestation of the larger "crusade," as Bush himself once called it. As W. J. T. Mitchell asserts, these images proliferated for a brief time with the rapidity and uncanny duplication of the act of cloning, and in each new manifestation they accreted meanings and interpretations along the way. In reference to the infamous "hooded man" photo, he writes: "If ever an image has been 'cloned' in the circuits of mass media, this one was, both in the sense of indefinite duplication, and in the further sense of taking on a 'life of its own' that eludes and even reverses the intentions of its producers."[66] For Mitchell, the image's resemblance to Christian passion iconography and its transposition onto an Arab body indicate its inherent openness to interpretation in multiple pro- and antiwar discourses.

I would instead argue that the fluid nature of the images as a *collection* allowed them to be inserted into multiple competing discourses. That is, lacking a fixed story of their own, the database of images from Abu Ghraib provided ready source material for people on every side of the issues involved: when the images emerged, they had no captions to anchor or interpret their meanings. As Morris, echoing Susan Sontag, claimed in his first blog post, "[A] picture unaccompanied by words may not mean anything at all."[67] But as they circulated through the mediascape, any number of commentators stepped in to fill the void. Thus, the same "hooded man" image appeared on Fox News with the caption "Detainee 'Abuse'" and on the cover of *The Economist* with the headline "Resign, Rumsfeld."[68] This is exactly the flexibility of meaning enabled by the database, and it is this aspect of the Abu Ghraib images that the film repeatedly highlights in its CGI sequences depicting them moving about the screen.

And this is why the film generated so much controversy among critics and academics and so little interest among viewers. That is, by opening up these images to multiple interpretations and by insisting, as Morris's films always do, that the images themselves mean nothing outside of a specific discursive context, the film confronted a sociopolitical landscape already heavily populated with very definitive interpretations. And unlike McNamara's reflection on events and debates over thirty years old, these discourses were still in wide circulation. Coming rather late to the party, the film's claim that these images are still open to reconfiguration proved to be an unwelcome contribution to the discussion. Documentaries, after all, are interpretations of the historical world that invite us to agree or disagree—a move Bill Nichols describes with the enjoinder "This is so, isn't it?" Unfortunately for Morris and his studio, a majority of the viewing public answered this question with a resounding no.

Outside of the film's success or failure, however, considered among his other projects of the last decade it clearly stands as his most technologically driven project to date in both form and content. As digital media came to dominate the field of filmmaking, its implications for truth and representation obviously came to dominate Morris's projects as well. But an event as divisive as the Abu Ghraib scandal lacks the historical distance and twenty-twenty hindsight that the Crimean or even the Vietnam War provides, and as a consequence, the film became swept up in the controversy it explored. While the film has already faded into the background of documentaries addressing torture, it nonetheless exemplifies the ever-growing integration of moving images and digital media, making its thoughts on the dangers therein all the more timely.

CONCLUSION

If we return to the rather cryptic epigraph at the beginning of this chapter that first adorned errolmorris.com in 2000, it now seems a prophetic inauguration for the shifts to come in both Morris's own work and the political landscape we have been considering here as a whole. As Morris's interests expanded from film to many forms of media both old and new, Chesterton's fear of the "labyrinth that has no center" seems to have served less as a warning than as an inspiration for Morris. Amid a decade that witnessed the extreme polarization of American politics, however—an era in which the center all but disappeared—Morris's work seems to have heeded the call. Moving from the more arcane fringes of obscure Americana into the stormy waters of political filmmaking, Morris's two major projects from this period offer unique attempts to carve out an ethical and political center in the issues they explore. Surely Morris had made political films before, but nothing on this scale.[69] Taking on such notorious figures as Robert McNamara and Lynndie England guarantees that viewers will come to these films with strong, preconceived notions about their subjects—notions that the films attempt to confuse rather than clarify. Thrusting us into the center of complicated ethical issues, both films further force us to empathize to some extent with those who were vilified as the "bad apples" or "bad guys" in their respective circumstances—a move that McNamara himself reminds us is essential if reconciliation is to be achieved and humanity preserved. If his next project was any indication, Morris's work from this decade may prove to be an aberration. Released in 2010, *Tabloid* centers on former Miss Wyoming Joyce McKinney and her odyssey of kidnapping, cults, and nonconsensual sex. While it certainly continues his preoccupation with mass media and social mediation, Morris has returned once again to his previous emphasis on idiosyncratic subjects. While it may prove to have been a detour for Morris, however, the intersection of film, politics, and technology that these two films explore was rapidly becoming the center of online activism and documentary filmmaking during the early presidency of George W. Bush.

3

Networked Audiences

MoveOn.org and Brave New Films

Revolution doesn't happen when society adopts new technology, it happens when society adopts new behaviors.

—CLAY SHIRKY, "*HERE COMES EVERYBODY*"

On December 4, 2016, a man carrying an AR-15 stormed into Comet Ping Pong, a pizzeria in Washington, D.C., and demanded to see evidence of the child sex-trafficking operation that he believed was headquartered in the basement. Over the preceding months, stories had been circulating on InfoWars and various other right-wing news websites about the alleged conspiracy and its connections deep within the Democratic Party. Several mainstream news organizations including the *New York Times* and the BBC had covered and debunked the story, but prominent Republicans in the Trump transition team continued to fuel speculation on Twitter, and the man had the impression that "something nefarious was happening."[1] Though no one was injured, "Pizzagate" set off an immediate series of alarm bells about the power of fake news to mislead people, and the role of social media in accelerating its spread. Alongside the growing awareness that similar "news" sources might have helped Trump win the election (a topic addressed more fully in chapter 6), the incident seemed symptomatic of a much wider ailment within the media and the public. But long before the 2016 election, before Hillary Clinton was a candidate for office or Facebook a website, independent sources on the left were decrying what they described as right-wing media manipulation. The culprit was the cable network Fox News, and its accusers were MoveOn.org and Brave New Films, a pair of progressive grassroots media organizations working to connect and galvanize members of the left.

Independent media production has a deep history of both working to effect political change and critiquing more-established media in the process. In the title of the introduction to his influential study of radical political documentary on the left, *Show Us Life,* Tom Waugh cleverly poses the inversion "Why Documentary

Filmmakers Keep Trying to Change the World, or Why People Changing the World Keep Making Documentaries."[2] Like the book it is drawn from, the title hits directly on a theme that has run throughout the last eighty years of documentary filmmaking—namely, its connection with the people and organizations that hope to produce social change. Waugh's title also uncovers two possible routes to the production of a social-issue documentary: the first stems from the desire to "change the world" and settles upon documentary as a means, while the second originates in the ability to make a film and alights on a particular issue as an application of the medium. Either way, Waugh's playful rearticulation firmly binds political activism and social change with documentary film. But it also inadvertently describes the paths of two progressive activist units—the husband-and-wife team of Wes Boyd and Joan Blades, and the filmmaker Robert Greenwald—whose respective organizations, MoveOn.org and Brave New Films, would take dramatically different routes toward the level of hybridity that Waugh implies. What Waugh couldn't have foreseen in the pre-Internet era in which he was writing, however, was the importance of newly available digital technology for both approaches.

The political polarization that intensified after 9/11 radicalized a new generation of political activists who already possessed lives and livelihoods outside of organized party politics but who nonetheless felt called upon by the events they saw unfolding to *do something* about newly perceived injustices. A figure like Jon Stewart, for example, combined a career in comedy and entertainment with an impulse to speak out politically into a new form of political entertainment, *The Daily Show*. This found an audience among a generation of like-minded and similarly politicized viewers.[3] For Blades and Boyd, and for Greenwald, this metamorphosis took the form of blending careers in technology and in filmmaking, respectively, with large-scale political organizing to create two of the most influential independent political organizations to emerge during the decade after 9/11. As of the 2016 presidential election, MoveOn.org boasted over eight million members and participated daily in organizing campaigns across the country on targeted issues from civil rights to health care to budget reform.[4] For its part, Brave New Films was responsible for some of the decade's most successful and influential political documentaries, from *Walmart: The High Cost of Low Prices* (2005) and *Outfoxed: Rupert Murdoch's War on Journalism* (2004) to several of the most overt attacks on the Bush administration and its war policies, including the so-called "Un" Trilogy—*Unprecedented* (2002), *Unconstitutional* (2004), and *Uncovered* (2004)— and *Iraq for Sale* (2006), among others.

After several years of collaboration, the two organizations evolved into surprisingly similar operations. In spite of the early (and in many cases pioneering) social media that enabled it to become an archetype of netroots organizing, MoveOn.org regularly turned to the decades-old technology of documentary media as a means to mobilize members. Likewise, Brave New Films began to rely heavily on social media in order to fund, publicize, and even produce its film projects. Both ended

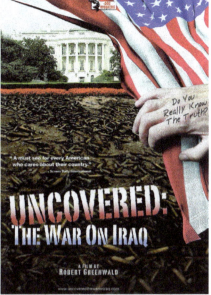

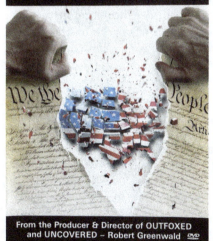

FIGURE 3.1. Robert Greenwald's "Un" Trilogy marked his move into political documentary and brought him into collaboration with MoveOn.

the decade as hybrids of documentary film and political activism of the sort that Waugh alluded to twenty years earlier. For both, however, the glue that enabled this synthesis was the technology that had emerged in the two decades since. Their parallel evolution, moreover, is not simply a coincidence. At several points around key political events during the Bush years—notably, the 2004 elections—the two organizations collaborated on projects that convinced each of the efficacy of the other's tactics. Both organizations started the decade with the conviction that their respective media forms (filmmaking and the Internet) could, as Waugh put it, "change the world," but both left the decade with the conviction that it would take some combination of both to do so.

This chapter looks at these organizations and considers the way both utilized differing measures of documentary film and digital media to change the world. While scholars and historians look to the role of YouTube videos in the Iranian Green Movement in 2009–10 or the part played by social media technology like Facebook and Twitter in the Arab Spring or the rise of the alt-right over the last several years, Greenwald and MoveOn were pioneering similar practices years before these eventual mainstays of Web 2.0 even existed.[5] Despite their disconnected roots in technology and filmmaking, during the period of 2000–2008 the two organizations collaborated with one another, and both evolved into hybrid organizations that challenge easy distinctions between documentary film, political activism, and social media. The heated, gloves-off environment of political debate surrounding the series of close political contests from 2002 through 2008 emboldened both groups to engage in radical media experimentation to advance their political agendas. Their individual and shared histories during this period provide an ideal illustration of the natural synergy between these forms. In MoveOn and Greenwald, we find a praxis-driven example from early in the era of social media that reflects many of the broader theoretical debates that would eventually emerge. As Greenwald and MoveOn both demonstrate, people trying to change the world were still making documentary films, but they were also doing other things as documentary images became one part of a widespread strategy aimed at social change.

MOVING IN THE SAME DIRECTION

The MoveOn–Brave New Films collaboration begins with the enormous popularity of two unrelated pop-culture relics of the past: singer, songwriter, and sometimes actress Olivia Newton-John; and the iconic Flying Toasters screen saver of the pre-Internet computer. Both were the forerunners, and in a sense the angel investors, of what would later become Brave New Films and MoveOn.org. Before his engagement with political documentary, Robert Greenwald worked for several decades producing and directing what the *New York Times* described as "a number of commercially respectable B-list movies,"[6] including 1984's *The Burning Bed*, starring Farrah Fawcett, and 1980's *Xanadu*, starring Olivia Newton-John. While several of these early films evince a clear interest in social issues, nothing foreshadows the dramatic transition Greenwald made in the wake of the 2000 presidential election to producing and directing some of the most critically and commercially successful political documentaries of the post-9/11 period.

Joan Blades and Wes Boyd got their start founding Berkeley Systems, a San Francisco Bay Area software company that created a number of different applications for the Mac including the early text-to-speech program Outspoken and the virtual-desktop program Stepping Out. Mainstream success arrived for the company with its popular screen-saver program After Dark, which featured the signature Flying Toasters, and the later trivia game You Don't Know Jack. After selling the company in 1997, Blades and Boyd began circulating an online petition via e-mail in the wake of the Monica Lewinsky scandal that directed Congress to "censure President Clinton and move on." The petition eventually generated over a half a million signatures and established an issue-oriented, technology-driven campaign model that the political action group has followed ever since. Since its founding, MoveOn has experimented with and adopted various social media technologies like Meetup, Facebook, and Twitter to expand and extend its network of political activists. It has covered a number of domains ranging from election campaigns for individual candidates to more general issues like health-care reform, gun control, and the wars in Iraq and Afghanistan.

As innovators in the fields of film production/distribution and political organization, respectively, Brave New Films and MoveOn have both been the objects of extensive study by film scholars, political scientists, and sociologists seeking to analyze the impact they have had in producing social change. Charles Musser and Christian Christensen, for example, both point out the innovative distribution techniques Greenwald and Brave New Films pioneered in the period from 2004 through 2008.[7] Christensen demonstrates that Brave New Films, via its partner organization Brave New Theaters, began building nontraditional screening outlets for its films (in homes, churches, and other public venues) into a hub for sympathetic groups and individuals to initiate further social action.[8] Similarly, MoveOn has been the object of extensive research for social and political scientists seeking to unpack the group's use of newly evolving technology for political organization and mobilization. Studies have focused on the role of MoveOn in relation to other grassroots movements, the group's use of technology (particularly e-mail and other social media) to create a new model for social movement organizations, and the rhetoric of its campaign materials in manufacturing a virtual imaginary community.[9] In addition to this, both organizations have received an impressive degree of attention from the mainstream press.[10]

Less discussed, however, has been the influence of the two organizations on one another—a critical oversight given the influence of older media practices like documentary activism on new technologies like social media. Furthermore, despite an acknowledgment of the obvious role of films and other media in their efforts, most of the coverage has left aside any formal or aesthetic discussion of specific pieces of media (films, videos, e-mails, etc.) and what role these qualities might play in shaping the tone or direction of the action to be taken. What follows will argue for the essential role that the documentary form played in fostering a new model for media activism and political participation in the post-9/11 period.

ROBERT GREENWALD:
FROM *XANADU* TO AFGHANISTAN

Prior to working with MoveOn.org, Greenwald got his start in the documentary form when Richard Ray Perez and Joan Sekler, both longtime activist filmmakers, approached him with "paper bags filled with tapes they had shot in Florida."[11] Outraged by the outcome of the 2000 election, Greenwald found it surprising that no one else was working on a film about the myriad of controversies and inconsistencies surrounding the Bush victory. He agreed to take on producing the project, which eventually became *Unprecedented: The 2000 Election* (2002), directed and cowritten by Perez and Sekler. Timed to coincide with the 2002 midterm elections, the film premiered on September 17, 2002, and went on to be screened at several film festivals and high-profile events but did little to change the results of the election, which was widely perceived as another victory for the right. Its impact on Greenwald, however, was significant. As Musser put it, although "the documentary changed the trajectory of Greenwald's filmmaking career . . . its limited distribution and impact provided the filmmaker with issues to ponder as he looked toward the 2004 election."[12] While he felt confident that his films were focused on the right issues, he wasn't sure that the right people were seeing them.

In June of 2003, with the war in Iraq already well under way, Greenwald began work on his next film. As with *Unprecedented*, his goal was to shed light on an issue being ignored by the mainstream media. For Greenwald, the film seems to have resulted from something of an epiphany:

> It was an early morning in late June, I was reading the paper, and in the middle of a long article about Iraq, one of the Bush administration folks was quoted, speaking about "programs for weapons of mass destruction" and how sure he was that they would find "programs." I got a knot in my stomach and a feeling of deep concern. We did not go to war for a program. . . . We went to war because we were told there were "weapons" and that the threat was imminent and dangerous. But the article did not in any way challenge this revisionist explanation of the "why." I imagined a headline—'Programs for WMD Found!'—and I feared that we would all just accept that.[13]

This recollection reveals that his decision to make the film stemmed from two interrelated forces: his perception that the administration had changed tactics and, equally significant, his conviction that the mainstream media was failing to hold it accountable. His oppositional stance toward both institutions (the government and the media) not only informed the overall direction of his career afterward but also placed him in step with a number of other newcomers to the progressive-media landscape, including comedians like Jon Stewart and Stephen Colbert, emerging left-wing bloggers on sites like the Daily Kos and Huffington Post, and of course, MoveOn.org. This group of activists, forged in the polarized environment of post-9/11 politics and empowered by new digital frameworks, formed the cohort that Theodore Hamm referred to as "the new blue media."[14] Although still a novice to documentary film, Greenwald was immediately drawn to its potential as an alternative to the mainstream media. This same potential had been attracting activists and artists, from the Workers Film and Photo League in the early 1930s through groups like Newsreel in the 1960s and on to the groundbreaking (and ongoing) efforts of groups like Paper Tiger Television and Deep Dish TV to sidestep the mainstream media.[15]

Like these forerunners, Greenwald realized that an effective alternative media required not just a different message but also a different channel of distribution. He states: "I have made over fifty films including theatrical, cable, and television, all utilizing the existing distribution system. In the case of *Uncovered,* I wanted it seen quickly. So I never considered the traditional gatekeepers."[16] For Greenwald, circumventing the existing distribution system entailed approaching John Podesta of the Center for American Progress (CAP) (a newly formed progressive think tank largely funded by George Soros) and Wes Boyd of MoveOn. Both organizations provided funding for the completion of the film but also, and perhaps more importantly, tapped into their existing member networks for what Greenwald called an "upstairs-downstairs" distribution model. This involved CAP organizing screenings for key decision makers (including every member of the House of Representatives and the Senate—the presumed "upstairs" center of power) and MoveOn organizing screenings in twenty-six hundred house parties across the country (the "downstairs" segment of disaffected voters.)[17] Further upending the traditional distribution model, Greenwald also sold DVDs of the film directly from his website via alternative outlets beyond CAP and MoveOn, including AlterNet, BuzzFlash, and *The Nation,* eventually enlisting a commercial distributor and selling over 120,000 copies of the film. As its reputation grew, the film attracted the attention of a commercial distributor who took it to the Cannes Film Festival and released a longer version in theaters around the world.

In addition to breaking new ground in distribution, *Uncovered: The Whole Truth about the Iraq War,* also forged another principle Greenwald's films have adhered to since: timeliness. Initially, Greenwald had planned on a year to complete the film, but at the request of Wes Boyd (who asked if it would be possible to complete it in a month), Greenwald cut the schedule down to just under five months. While certainly longer than the immediacy of mainstream television news coverage, by film standards this time frame is relatively quick. As events continued to unfold and new information came to light, Greenwald further demonstrated a willingness to rework the film, eventually expanding it by nearly thirty minutes and shortening the title to *Uncovered: The War on Iraq*. Later, even five months would seem too long. Greenwald's 2008 project *Rethink Afghanistan* was shot in a series of installments that were released to the web before eventually being re-worked into a longer DVD release for event-based screening and direct sales.[18] Greenwald's desire for speed and a more flexible cinematic text demonstrate both his desire to compensate for the poor coverage of the war by the mainstream news media and a wholesale reliance on emergent technology in order to do so.[19]

With *Uncovered*, Greenwald established the two features that have been the hallmarks of his activity since: (1) cooperation with other activist groups for production, funding, and exhibition (what Christian Christensen identifies as the "coalition model" of documentary advocacy[20]); and (2) distribution via whatever technology will allow the work to be seen by the greatest number of people in the shortest amount of time, be it DVD, theatrical release, or, eventually, online streaming. The film's commercial and critical success firmly established Greenwald in the circuit of progressive liberal activists and media makers—connections he would increasingly rely upon in future projects. Greenwald's next project not only perfected this model but also resulted in a newly formed production company–cum–activist organization, Brave New Films, which has since become the umbrella organization for all of his political activities. But before Greenwald could take that next step, he needed some additional help from MoveOn.org, which itself was quickly evolving from an e-mail petition to a political media powerhouse.

MOVEON.ORG

The story of MoveOn.org's evolution toward political power and media advocacy offers a paradigmatic example of the "power of the Internet" that has now be-come commonplace, one in which an organization's speed of success comes as a surprise for everyone involved, including its founders. Although the unexpected is by definition difficult to anticipate, MoveOn managed to capture that spirit re-peatedly in its first decade of existence. The viral success of Boyd and Blades's original e-mail petition to "censure President Clinton and move on" (garnering hundreds of thousands of signatures in a few weeks) exemplifies an often repeated theme in media accounts of its organizing ability: an ability to capitalize on pub-lic reaction by raising money quickly or turning out supporters for last-minute events. Although MoveOn made an early push for tougher gun legislation in the wake of the Columbine High School shootings in 1999, for the most part its early years were focused on issues related to the Clinton impeachment and reshap-ing Congress away from its Gingrich-based social conservatism.[21] In the run-up to the 2000 election, MoveOn repeatedly broke online fund-raising records for candidates it supported in races against some of the most outspoken proponents of impeachment, including James Rogan of California (the House impeachment manager) and Florida Congressman Mark Foley.[22] Although it scored a few early victories in these races and established itself as a player in political fund-raising and viral campaigning, MoveOn sat out the postelection protests over Bush's election (a move Boyd later regretted) and seemed resigned to periods of inactivity between election cycles.

However, the terrorist attacks in New York and Washington, D.C., on September 11, 2001, and their political consequences pushed MoveOn's membership and its founders to a more issue-oriented protest model. Rather than focus solely on elections, MoveOn began mobilizing between elections to oppose specific policies. In 2001, the group merged efforts with 911-peace.org and recruited its founder, Eli Pariser, to be its executive director. In a story reminiscent of MoveOn's own, on September 12, 2001, Pariser had sent out an e-mail to thirty friends asking them to sign a virtual petition he set up urging "moderation and restraint" in response to the attacks. In two weeks, the petition generated over five hundred thousand signatures and elevated the twenty-year-old Pariser to national prominence as a leader in the growing protest movement to the invasion of Afghanistan. It also brought him to the attention of Boyd and Blades, who clearly recognized Pariser's potential and saw him as a natural fit for their efforts.

Over the next few years, Pariser would be instrumental in MoveOn's foray into media campaigning as a key component of its political strategy. In early 2002, during the buildup to the war in Iraq, the group launched another online petition calling on Congress to "let the [weapons] inspections work" and sought member donations to raise $40,000 for a full-page ad in the *New York Times*. When the effort generated nearly $400,000, MoveOn took this as a sign that its members were "very interested in being heard through advertising," as Blades put it.[23] MoveOn used the additional funds to create what became known as its "Daisy" ad, named after the controversial Lyndon Johnson television advertisement that aired during his 1964 race against Barry Goldwater. While reaction to the MoveOn version was mixed, it succeeded in generating attention and airplay far beyond the original thirteen cities in which it was shown as a paid spot.[24] David Fenton, MoveOn's communications consultant, claimed that thanks to its coverage on the Internet and cable news outlets, it had become the most viewed advertisement in the history of the medium.[25]

Seeking to build on this success, MoveOn next created a contest to replicate the success of the Daisy ad in a more distributed fashion. Called "Bush in 30 Seconds," it challenged MoveOn members to create a political ad that summed up the Bush administration in thirty seconds. The winner's entry would be aired during the Super Bowl halftime show, its broadcast paid for by MoveOn contributors. The contest was judged by a panel of celebrities, from the musician Moby (who was credited as one of the contest's creators) to other high-profile personalities

FIGURE 3.2. MoveOn's remake of the "Daisy" ad generated controversy and coverage, demonstrating the power of media for online organizing.

like Jack Black, Russell Simmons, and Michael Moore.[26] Citing a policy against advocacy advertising during the game, CBS declined to sell MoveOn the spot. But the winning ad and the controversy the contest generated nonetheless earned an enormous amount of free publicity for MoveOn.[27] MoveOn continued to make political advertising a primary tool in its efforts throughout the next few years, spending over $10 million airing its own material in the 2004 election alone.[28]

A closer look at the home page for MoveOn from January 2004 (available via the Internet Archive's invaluable Wayback Machine) demonstrates the extent to which MoveOn at this point already conceived of politics and media—both new and old media—as an intertwined enterprise.[29] Laid out in a standard three-column format with a header and footer at the top and bottom, the page essentially remediates the format of a newspaper—or, in the case of MoveOn, perhaps a newsletter or pamphlet is the more relevant print reference.[30] Red, white, and blue predictably dominate the color scheme, implicitly emphasizing that this is a newsletter about the state of US politics and democracy. The top-level categories across the top announce to visitors the organization's areas of focus, from informational ("Home" and "About"), to referential ("Press Room" and "Media Coverage"), to political praxis ("Make a Donation" and "Become a Volunteer"). While they appear as discrete categories, however, these different areas—information, action, organization—are connected as equal parts, or steps, in a cohesive whole.

FIGURE 3.3. The MoveOn home page circa January 2004.

"Democracy in Action," the header claims, involves pulling individual activity and resources (time and money) into a collective, organizational form (MoveOn itself) in order to provide information and influence the larger mediasphere.

This theory of politics carries across the lower, content portion of the page, finding various iterations in each of the specific stories and items featured. The organizational logic behind the page seems to devote the left-hand sidebar to past items, each with an image; the center frame to the current campaigns and information; and the right-hand sidebar to quicker, press-release style bulletins— almost lending the whole a "past, present, and future" split. In its content, each of these items cements MoveOn's larger message. On the left-hand side, links and images are given for three discrete streaming or rich media pieces, including a replay of the "Bush in 30 Seconds" winner, a recording of a MoveOn-sponsored lecture given by Al Gore on climate change (a full two years before *An Inconvenient Truth* in 2006), and an interactive map of the screenings and house parties that were held for the premiere of *Uncovered*. The central frame is dedicated to mobilizing members to sign petitions censuring President Bush for misleading the public and censuring CBS (the network carrying the Super Bowl) for boycotting its ads. Political impact, these central items imply, arises out of the regulation of the flow of information: providing information to supporters to recruit and mobilize them, creating information to convert others, and gaming the mass media into supporting these and other efforts.

The footer of the page further underscores and revisits these connections. The left side of the footer is dedicated to recruiting for MoveOn's "Media Corps" and reporting on its actions on the organization's behalf, and the right footer makes one last plea to members to join or donate. The Media Corps was a group of volunteers MoveOn had mobilized as a type of rapid-response unit focused on the mainstream media and its coverage of the war. On the recruitment page, it describes the Media Corps as a group of "committed MoveOn volunteers who will mobilize to push the media to fairly cover this war."[31] Volunteers were asked to commit to taking an "action" every day, which would usually involve contacting mainstream media outlets regarding their coverage of various issues. In its focus on shaping and critiquing the mainstream news by mobilizing its volunteers, the Media Corps offers an interesting precursor for the crowdsourced production model that Greenwald's *Outfoxed* would rely on later that year.

A great deal of MoveOn's resources (both the labor of its volunteers and the money that it collected from volunteers) were hence dedicated to influencing, making, and distributing media. Starting in 2004, MoveOn's other channel for distributing these short political advertisements has been through embedded video clips on its website and Facebook pages. While MoveOn had used streaming video in campaigns before (notably for the "Bush in 30 Seconds" contest), the advent of YouTube in 2005 brought simplified video streaming to mainstream users, obviating the need for custom browser plugins or software downloads to deliver video over the web to a mass audience.[32] Less than a year after YouTube launched, MoveOn had established a profile on the site and began uploading campaign-related videos to embed in their webpages. Short videos explaining the issue at hand became a regular feature on its campaign pages, alongside a brief written explanation and a form to use to respond (donating, signing a petition, e-mailing a specific politician, etc.). Unlike the messages it paid to broadcast on mainstream television channels, these embedded campaign videos were more akin to in-house advertising and offered short bursts of information and rhetorical appeals to incite the viewer/member to some kind of action. Since its first posts, MoveOn has posted hundreds of these short videos on its YouTube channel, which, collectively, have been viewed millions of times. YouTube's allowance of embedded video on other sites meant that a significant portion of these videos were watched by users visiting the MoveOn campaign pages rather than on YouTube's site. Most users would have encountered them within the context of an overt political message rather than in the heterogenous context of user-submitted video that forms the bulk of YouTube's content. While the inclusion of streaming video on a webpage has by now become commonplace across the Internet, MoveOn's specific use and early experimentation were a clear indicator that the organization realized the power of moving-image media in general.

Beyond direct advertisement and short embedded video clips, MoveOn's most consistent use of media in its campaigns has been its support of outside projects it feels are relevant to its larger goals. After the dispute between Michael Moore and the Disney Corporation over the distribution of Moore's film *Fahrenheit 9/11* (2004), MoveOn started a pledge drive of members willing to see the movie on its opening weekend, hoping to make the film a success in spite of the efforts to block it. In exchange, Moore participated in an online virtual "town hall" meeting that connected thirty thousand members at a number of house parties across the country with the director to discuss issues raised in the film. In calling on members to see the film, Pariser praised MoveOn for taking up a mission similar to Greenwald's: holding the administration accountable on issues when the mainstream media didn't. He wrote: "Despite years of television coverage on Iraq and the war on terror, most of the movie consists of footage you'd never see on TV. . . . The film is filled with this stuff, and it's hard to imagine seeing it and not being moved, shocked, and outraged."[33] Since then, MoveOn has sponsored screening and attendance drives for many films, including *An Inconvenient Truth* (Davis Guggenheim, 2006), Moore's later films *Sicko* (2007) and *Capitalism: A Love Story* (2009), the Leonardo DiCaprio–produced and -narrated *The 11th Hour* (Leila Connors and Nadia Connors, 2007), the Iraq war film *The Ground Truth* (Patricia Foulkrod, 2006), and, of course, several of Robert Greenwald's films, including *Outfoxed: Rupert Murdoch's War on Journalism* (2004), which would prove to be the most extensive collaboration between the two organizations.

FOX NEWS: A COMMON ENEMY

In mid-2003, reports began surfacing in the mainstream press that a Republican-led effort to filibuster the Senate in an all-night session pushing for an up-or-down vote on George W. Bush's judicial appointees had in fact been the idea of an editorial in the *Weekly Standard,* a publication owned by Rupert Murdoch's News

Corporation.[34] Furthermore, the *Washington Post* and *The Hill* both reported that Fox News had asked Republicans to schedule the filibuster to coincide with the opening of *Special Report with Brit Hume* at exactly 6:02 p.m. (EST) to capture their dramatic entrance live on television.[35] While the story itself simply became a footnote, for progressive activists and media-watch groups, it seemed to provide clear evidence for their long-held suspicion that Murdoch's company, and Fox News in particular, were heavily biased in favor of Republicans. For MoveOn. org and Robert Greenwald, it was a call to arms—one that would direct their next collaboration and shift the future direction of both organizations.

Given their mutual opposition to the Bush administration and their mistrust of the mainstream media's ability to hold it accountable, Greenwald and MoveOn saw in Fox News an opportunity to critique both groups at once. Claiming, as Greenwald did, that "Fox is a Republican, not merely a conservative, network" meant that confronting the network would simultaneously allow them to confront the entire Republican agenda.[36] This move, in turn, further solidified the position of MoveOn as an alternative to conservative and Republican policies and Greenwald and other filmmakers as alternative media outlets to the mainstream press.

Fox had long been an object of scorn for the left based on what many saw as its destructive effect on television news in general. Initially dominated by CNN, the market for twenty-four-hour news began expanding in 1996 with the addition of Fox and MSNBC. To launch the new network, Murdoch hired Roger Ailes, a former NBC executive and Republican political consultant. Ailes was responsible for designing the network's emphasis on live news coverage during the day followed by opinion programming in the evening. To anchor these evening programs, he hired large personalities like Bill O'Reilly and Sean Hannity in order to differentiate Fox from the staid programming of CNN. The network's emphasis on visually dense graphic presentation and sensational stories earned it comparisons to *USA Today*. But despite these dismissals, between the terrorist attacks in 2001 and the run-up to the Iraq war, Fox moved into first place in the cable-news ratings—a prominence that drew attention to its effect both on cable news specifically and on American political opinions in general. The heightened tension surrounding the 9/11 attacks and an impending war played directly into the dramatic, sensational presentation that Fox brought to the business of television news. Its style was so extreme that even First Lady Laura Bush once criticized the network for "scaring people" with its continual coverage of the terror threat level.[37] As Ken Auletta pointed out in a widely read profile of Fox and Roger Ailes from 2003, CNN and MSNBC both found themselves in the position of playing catch-up, often by ineffectively imitating the leader.[38] Moreover, many widely suspected that conservative media such as talk radio and Fox News were responsible for the ascendancy of the Republican Party on a national scale—an influence sociologists would later dub "the Fox News factor" or "the Fox News effect."[39] Thus, for progressive political action groups like MoveOn, Fox News was not just a convenient target, but rather an essential one for the advancement of the progressive agenda.

In 2003, MoveOn and Greenwald teamed up again to work on a political action campaign that would expose what they believed to be Fox's abuse of mainstream journalism, the centerpiece of which would be the Greenwald documentary *Outfoxed: Rupert Murdoch's War on Journalism*. As with *Uncovered,* MoveOn provided production funding alongside the Center for American Progress, but this time it also lent its numbers and organizing strength to the production of the film itself. After looking over six months' worth of twenty-four-hour-per-day recordings of the Fox News channel, Greenwald outlined what he believed to be the most egregious of the Fox News tactics, which he categorized into a series of themes that would form the backbone of the film. MoveOn then put out a call to members asking for "Fox Monitors," individuals who would sign up to watch Fox News programs during specific times throughout the week.[40] When monitors found examples of Greenwald's themes, they would fill out a spreadsheet detailing the date, time, and context and forward it on to him. Greenwald then compiled these reports for his team of editors, who would pull the footage and work it into the appropriate sequences. To complement the Fox footage, Greenwald also conducted interviews with a series of former Fox employees, several of whom disguised their voice and appearance, as well as outspoken critics of the network like Al Franken and Eric Alterman.[41]

Once *Outfoxed* was complete, Greenwald, together with CAP and MoveOn, pursued the same upstairs-downstairs distribution method that they had used on *Uncovered.* To leverage the film, MoveOn conducted a series of specific actions based around the film that it dubbed the "Unfair and Unbalanced" campaign. These included a petition to the Federal Trade Commission to block's Fox's use of the phrase "Fair and Balanced" on the basis that it was inaccurate and misleading, a night of 2,750 house parties to screen the film, and a series of press releases and e-mail campaigns calling on Congress to force the network to "come clean about its rank partisanship," as Wes Boyd put it.[42] In his letter to MoveOn members urging them to participate in the campaign, Boyd also announced that members who made a thirty-dollar donation to the alternative news organization AlterNet would get a copy of the film for free, stating, "As part of this campaign, we've got to support good media, and AlterNet is a great independent outlet." In addition to AlterNet, MoveOn, Greenwald, and CAP also teamed up with several other independent news organizations and watchdog groups including FAIR (Fairness and Accuracy in Reporting), Common Cause, Media Matters, and the Center for Digital Democracy, among others.

Predictably, Fox responded to the claims presented in the film. Instead of suing for copyright infringement (as it had against Al Franken for his book *Lies (and the Lying Liars Who Tell Them): A Fair and Balanced Look at the Right)*, Fox instead leveled its own accusations on air against MoveOn, the *New York Times,* and George Soros for colluding to "corrupt the journalistic process."[43] Fox's insinuation that the *New York Times* was itself a liberal mouthpiece overtly echoes claims that the network, and the right in general, have repeatedly made about the general liberal bias of much of the mainstream media news, from NPR and CNN to the *New York Times.*[44] Regardless of the validity of either side's accusations—a topic too vast and vexing to take up here—these claims only further bolstered progressive calls for an independent media apart from the larger corporate conglomerates that had come to dominate virtually every channel of the news media. Indeed, the website for *Outfoxed* directs visitors to "sign the petition to break up the big media conglomerates and get higher quality news" and to "volunteer with Independent Media Centers all across the globe."[45] What Fox failed to realize in its counterattack was that the progressive activist groups aligned against it weren't advocating for one corporate media organization over another, but rather against corporate media organizations in general. For Greenwald, the entire shift of his career into documentary filmmaking and his partnership with organizations like MoveOn were based on the belief that people not only wanted but also needed an alternative form of media and a different channel through which to access it. While this desire would tangibly manifest itself in a series of mixed results with the rise of midstream media outlets such as *The Intercept* and *Breitbart News* (see chapter 6), in 2003 Fox News and corporate media were the problem.

OUTFOXED: THE EVENT

Like much of Greenwald's documentary output, *Outfoxed* works within a bare-bones style that offers little in the way of aesthetic flourish or formal innovation. The talking-head interviews juxtaposed with footage from Fox itself simply seek to prove the case, beyond a reasonable doubt, that the network engages in blatant political favoritism and consistently presents editorial opinion as unbiased news. As one reviewer described it, "[T]he result is an unwavering argument against Fox News that combines the leftist partisan vigor of a Michael Moore film with the sober tone and delivery of a PBS special."[46] The film's strongest evidence is its use of the Fox material itself to decontextualize and lay bare what it describes as the Fox style of journalism. Its smoking gun lies in a series of memos leaked to the filmmakers by Fox employees (and later published on the film's website) that detail the way Fox sets its agenda for covering specific events. The memos, sent out by Fox's vice president for news, John Moody, suggest specific angles from which the news should be approached, and are the film's irrefutable evidence that Fox directs its employees to cover the news from an overtly political standpoint.

The film's conclusion—that Fox News is politically biased—is, as the *New York Times* put it, "not exactly earth shattering."[47] By the time it was released, studies had already been published by the Columbia School of Journalism and Maryland's Program on International Policy demonstrating these exact conclusions, and Fox's particular style of reporting was sufficiently well known that it had already become fodder for parody on popular political satire outlets like *The Onion* and *The Daily Show*.[48] But the simple fact that the point had been made elsewhere doesn't make the film itself irrelevant, nor does it deny the utility it afforded the progressive groups who created it as a documentary. Although academic studies had been carried out and jokes had been delivered, all pointing to the same conclusion, the case had yet to be made in the particular form that *Outfoxed* delivered. Put differently, the "sober tone . . . of a PBS special" that the film manifests is not simply the result of a lack of creativity but, rather, an intentional part of its larger rhetorical strategy.

Form in this case refers not only to the particular organization of *Outfoxed* itself but also to the documentary form in general, both of which played a decisive role in the film's argument and in its circulation within progressive political discourse when it premiered. As a type of moving-image media, documentary naturally lends itself to the method of media quotation essential to the film's strategy of using Fox's own material against itself. As Julia Lesage points out, documentary film is adept at taking vast quantities of information and synthesizing them down into salient points and a digestible format—a skill that would be brought entirely to bear by Greenwald and MoveOn as they sought to distill six months' worth of twenty-four-hour-per-day Fox News coverage down into the hour-and-a-half running time of the film.[49] This level of decontextualization and synthesis allows the film to achieve the small degree of stone-faced humor it allows itself as, for example, Bill O'Reilly repeatedly commands his guests to "shut up" in a variety of ways across a multiplicity of contexts, all presented in rapid montage succession. Pairing this sort of comedic evidence with the more traditional expert-interview segments positions the film directly between the academic studies and the parodic attacks that came before it.

Moreover, set in the context of mainstream-media criticism, the film utilizes the documentary mode to reflexively position itself as the type of product that it is advocating for. When paired with its unique production methods, *Outfoxed* becomes a powerful organizing tool for MoveOn and its members. In collaborating with MoveOn members to identify the material that would later appear in the film, Greenwald was effectively engaging in an early form of what Jeff Howe would later call crowdsourcing, or using technology to assign a large task to a group of disconnected individuals to complete in parts.[50] Crowdsourcing as a method of distributed labor would eventually come to refer to everything from the creation of Wikipedia to attempts by businesses to utilize slave labor in the developing world via platforms like Amazon's Mechanical Turk. The service, as Lily Irani and others have demonstrated, enables companies to quickly and inexpensively offload tasks that might otherwise require significant investment through a process that she refers to as "massively mediated microlabor."[51] As Irani persuasively demonstrates, the service excels at the "redistribution of tedium" and enables many information-era high-tech companies to maintain an aura of innovation despite fostering industrial-era levels of alienation and exploitation for the workers who create this value. While various tools and platforms have emerged to facilitate labor organization among the workers on these platforms, for the most part the crowdsourcing platforms themselves are antipolitical and anticommunal.[52]

But for MoveOn, crowdsourced production represented a unique opportunity to engage its members in the form of direct, participatory democracy that the group stands for. The implicit belief here is that a strong democracy requires an equally strong press to hold the government accountable and provide the electorate with the information they require to make informed choices. This is what motivated the attack on Fox and the mainstream media, and is overtly proclaimed in the rhetoric used by the independent media partners the film promotes as an alternative. For Greenwald and MoveOn, the first step in achieving this goal was exposing what it felt was the most egregious example of corrupt journalism—a goal it could achieve only through the help of its many members. In this sense the film is a tangible product of group action, much the same way that MoveOn tries to be.

The documentary form also lent itself to a particular method of organizing that MoveOn had pioneered: the house party. If MoveOn members were an important part of the film's production process, they made up an even larger part of its intended audience. Through its use of the house party to screen films for its members, MoveOn had essentially become an ad hoc exhibitor for political documentary films, able to produce thousands of viewers for a given film. Indeed, MoveOn's house-party event for Greenwald's *Walmart: The High Cost of Low Prices* put together seven thousand simultaneous screenings. Arguably, it enjoyed a wider release than James Cameron's *Avatar* (2009), which, at its height, played in 3,461 theaters simultaneously.[53] There is no comparison between a nationwide theatrical release sustained over many weeks and a stand-alone event arranged in a myriad of private venues, but in the realm of documentary film this level of exposure is impressive. Given Greenwald's commitment to political action, access to a group of committed audience members presented an enormous opportunity. Using MoveOn's member base would allow Greenwald to directly translate his film into action, not to mention DVD sales. (Party hosts are often required to purchase the films at their own expense, typically at a discount.)

For MoveOn, the events also double as membership-recruitment and volunteer drives. Since house parties are typically paired with a conference call afterward that includes MoveOn organizers and the filmmakers, attendees feel part of something larger than their particular gathering. As an eighty-six-minute text delivered in a "sober tone," a documentary like *Outfoxed* functions as a crash course for MoveOn members in the issues the organization seeks to advance. Of course, such events have long been a staple of political documentary films, often screened in public venues to sympathetic audiences to generate awareness of and motivation for specific issues or causes.[54] Where MoveOn differs from its forerunners is in the sheer scale it achieves through technology. When a house party is planned, members are notified via e-mail and other channels of the upcoming event and asked to host a party. Once a sufficient number sign up on the website to host an event, a second wave of e-mails invites others to RSVP to screenings in their area using their zip codes. If no party is being held nearby, members are asked to hold one of their own, soliciting participation from friends and family who might be interested. From here, hosts are asked to purchase copies of the film and provide an Internet-connected computer or telephone to participate in the discussion afterward. After the screening, hosts will typically call or log in to a conference call where MoveOn staff and/or the filmmakers will address the audience and answer questions.

Paired with this online connection is the offline, face-to-face interaction at the party itself. Situated in people's homes but open to the general public, these events hinge upon turning the shared experience of the film into an interpersonal connection among people whose primary interface with the organization that brought them together is through their computers. Reportage on the house parties in the mainstream press paints a heterogeneous picture that thwarts easy characterization, with accounts ranging from catered gatherings in the homes of wealthy celebrities attended by hundreds to small get-togethers in apartments offering homemade appetizers to a dozen people.[55] Regardless of the nature of individual events, however, in their mass simultaneity they offer a unique hybrid of public and private spectatorship that troubles the traditional distinction between the theater and the home. Despite talk of the Internet and the high-definition living room atomizing the traditional theater audience, the MoveOn house party works toward creating the same sense of collective spectatorship offered in a theatrical screening.[56] While hosts and other attendees are encouraged to publicize the screenings to friends and family as a distributed publicity tool, there is little expectation that a given event will be composed solely of people familiar to the host. The inclusion of the conference call also requires the nationwide screenings to take place at the same time. This simultaneity strengthens the collective sense of a shared experience, which in turn fosters greater identification with the organization and the collection of individuals who constitute it.

The structure of the house party also assumes that people will want to discuss the film and the issues it raises after they have watched it, and the gathering provides a natural outlet for the impulse. Centering the party on the film screening gives the people in attendance a built-in conversation topic on an issue they are already predisposed to support. Furthermore, it also gives them a chance to act upon the problem structure that organizes most of Greenwald's films. In spite of his reluctance to utilize the pathos-driven approach that comes more naturally to someone like Michael Moore, his films nonetheless deal with controversial issues that he and MoveOn feel people will act on. Attendees are often asked to conclude the event by writing postcards, signing petitions, making phone calls, or volunteering to participate in future events—all of which provide an outlet to respond to the call to action that problem documentary films typically end with. This move implicitly pairs collective spectatorship with collective action, and in the context of the house party, the two sit side by side. In this sense, the documentary form offers MoveOn the ability to bring its members together around an event (the screening), provide them with a shared body of knowledge (the film), and then ask them to act immediately on the information they have been given.

In the context of the house party, the "sober tone" of a film like *Outfoxed* thus sits ideally between the seriousness of academic studies and the accessibility of the popular-culture parodies on Fox's bias that had already been presented. The minimal style that Greenwald's films employ fits within a logic dictating that once people have the information, they will feel compelled to act. The house parties are structured to capitalize on this. The social ritual of coming together to watch a film and the impulse to participate in a nationwide "event" are both centered on the experience of the film text, and the film text derives its agency from the context in which it is presented. No other media form offers this same level of symbiosis.

OUTFOXED'S OFFSHOOTS

Greenwald himself went on to adapt the material from the film into additional forms of media—in this case, a book (cowritten with Alexandra Kitty) and a website (now defunct). Both offer interesting gestures toward the two ends of the spectrum that I am arguing his film sits astride. If *Outfoxed* the film is less weighty than more serious academic studies on the topic, then the book is clearly an attempt to make up the deficit. As was the case with the companion texts to both of Errol Morris's films from the same period, the book essentially works as a set of footnotes to the film: citing sources, reprinting transcripts from the interviews and Fox broadcasts, and presenting a more detailed argument than a film would allow.

And where the book grounds the film's assertions in a solid body of evidence, the website extended its argument and call to action forward in time and updated its documentation of Fox's tactics. In addition to a standard film website housed at www.outfoxed.com (which contained a trailer, synopsis, reviews, links to purchase the DVD, and other information), Greenwald also launched an offshoot site at http://foxattacks.com that contained a series of viral videos created after the film's release. Utilizing the tagline "They Distort, We Reply" in response to Fox's well-known "We Report, You Decide," the videos were intended to expose Fox's coverage of ongoing issues for inaccuracy and political bias. Thus when Fox went

FIGURE 3.4. The FoxAttacks home page circa October 2007. The "Pssst!" function on each video enabled users to share the video with others.

"on attack" against an individual or issue like Michelle Obama or health care, the staff at Brave New Films would post a video documenting their claims and notify subscribers, who would then forward the video on to others. For content, the site relied on partnerships with other independent watchdog groups like Newshounds and FAIR to alert it to inaccuracies in Fox reporting (a group it called the "Fox Attacks Coalition").[57] In essence, Brave New Films was creating a version of FactCheck.org or Politifact a decade before Donald Trump's eccentric campaign style made fact-checking news sites and political candidates a virtual necessity.

In another context, a website like FoxAttacks.com could be interpreted as an advertising mechanism to spur on DVD sales of the film, much as the popular *Freakonomics* blog implicitly advertised for the books even as it provided ongoing analysis that the books didn't contain.[58] In this context, however, there are several indications that the site's purpose was more than purely commercial. The first is that Greenwald, under the Brave New Films production company that came out of *Outfoxed,* has offered all of his films, including *Outfoxed,* for free on YouTube, Google Video, Facebook, and other platforms. The aim was to get the film's larger message out regardless of the particular channel of distribution. The second is that Greenwald made all of the source material in the film, including the interviews he conducted, available for use by other filmmakers under a Creative Commons license. Like the FoxAttacks website, publishing the raw material treated the film as part of an ongoing project for further development, as opposed to a finished text. More broadly, Greenwald has claimed that, like many documentary filmmakers before him, his aim is not to make money from the film, but rather to advance his political agenda.[59] Funded by nonprofit foundations and promoted for free across a variety of channels, such films, as well as their patrons, find their payoff elsewhere. As Jonathan Kahana notes, "Such entities usually expect to generate cultural, political, or ideological, rather than financial, returns on their investments."[60] In this sense FoxAttacks was a political extension of, rather than commercial support for, the film that inspired it.

FoxAttacks also demonstrated Greenwald's keen awareness of the need to tailor the broader message to the individual media channel in which it appears. If the house parties acknowledge that a feature-length documentary lends itself to discussion and follow-up action, the website asserts that a video clip on a web page lends itself to brevity and sharing. In making the videos highly portable (they were capable of being e-mailed to viewers and linked to and embedded in other web pages), Greenwald is taking advantage of the interconnected, entirely transferable nature of digital media on the Internet. In making them brief, he is admitting that the attention span of the Internet audience is relatively short. It is telling, for instance, that even after *Outfoxed* was made available on the web for free, DVD sales continued coming in, owing perhaps to the preference of one medium over another depending on the length and tone of the message, although habits in this regard are hardly fixed.[61] As Musser points out, this awareness of the power of the short form on the web would become key to Greenwald's work during the 2006 and 2008 elections.[62] But regardless, the lessons learned during the *Outfoxed* project would have an ongoing influence on the political activities of both Greenwald and MoveOn.

OUTFOXED: THE AFTERMATH

Although Greenwald and MoveOn continued to collaborate on projects and campaigns and to mutually support each other's projects, their integration has never been as extensive as it was on the *Outfoxed* project. Nonetheless, the collaboration established a model combining the informative power of moving images with the organizing power of the Internet that both groups have continued to follow. With the meteoric rise of Facebook as the primary social media outlet, both groups have migrated a great deal of their content sharing and event organizing onto the platform.

For its part, MoveOn still supports films and filmmakers who share its causes through house parties and screening drives. As video streaming became simpler and more ubiquitous, MoveOn's campaign messages and other web pages increasingly featured short video clips of materials relevant to a particular issue. Its use of these materials runs the gamut from thirty-second attack ads aimed at television to short testimonials by MoveOn members about an issue or their experience with the organization. Like FoxAttacks, MoveOn's video strategy demonstrates a canny awareness of the Internet as a medium. Two particular styles seem to dominate their output, and both offer examples of the larger role of MoveOn's in-house media production.

The first type of video that MoveOn regularly produces is closely related in form and purpose to the television advertising the organization has sponsored. As the "Bush in 30 Seconds" contest, the infamous "Daisy" spot, and its equally infamous 2008 follow-up "Not Alex" demonstrate, MoveOn has clearly become adept at turning a controversial advertisement aired in a few states into a national conversation topic that replays in news segments on television and becomes virally distributed across the web on blogs and other social media sites. MoveOn's in-house advertising for its specific campaigns may be intended for a different audience and distributed in a more limited fashion, but it nonetheless functions in much the same way. Slightly longer than a thirty-second spot, these videos are still quite short, in keeping with the brevity principle for web-based video. They reach their conclusions quickly and seek to make one brief call to action, usually via the web page in which they appear.

In 2018, for example, MoveOn created a campaign to preemptively enlist members to join protest rallies should Donald Trump interfere in the independent investigation into Russian meddling in the 2016 election. As part of its enlistment drive, it produced a series of short videos calling on MoveOn members to be prepared and register to join future protests, which it then posted to its Facebook page. One video, *If Mueller Is Fired Here's What To Do,* makes its case by featuring short testimonials from other members, including Karla in Illinois, Mary in Texas, and Steve from California, all of whom plan to protest should Trump attempt to disrupt the investigation.[63] The video ran just under two minutes and garnered nearly a million views in twenty-four hours, demonstrating MoveOn's continued ability to draw an audience. Like the original e-mail petitions that Blades and Boyd created in 1998 and Eli Pariser sent around in 2001, the hope for these videos is that they will "go viral" and spread around the Internet via e-mails, links, and posts to social networking feeds. But failing this, these embedded clips and the enjoinder to "share them with your family and friends" clearly serve another purpose, since most of the people viewing them on MoveOn pages are already members. Not only does sharing the video by posting it to a Facebook page, e-mailing it to a friend, and so forth work to publicize the issue, but it also provides members doing so with the sense that they have done their part for the cause. MoveOn's conception of itself as a member-driven organization clearly depends not just on getting its members to participate but also on giving them the perception that this participation is what makes a difference. Providing them with a palatable, humorous, easily distributable message introducing an issue enables these clips to serve both functions.

The second type of video that seems to dominate MoveOn's video production is the member testimonial. These clips feature members, often self-recorded via webcams, testifying to their experience with MoveOn and entreating other members to become similarly involved. In another context such limited production values might be a detriment, but in this context the format actually works as a strength. In a video called *Host an Event!*, a woman named Elinor shares her experience hosting a party for MoveOn and asks others to follow her lead.[64] Sitting before the camera in what appears to be a dimly lit living room, Elinor states, "If you've only interacted with MoveOn on the Internet, let me tell you, it's better in real life!" She goes on to ask viewers to "[s]ign up now to deliver a petition to your congressman, it's an easy thing to do. All you have to do is fill out the form that's located there, or there, or maybe there." As she gestures around the screen, acknowledging her virtual location on a web page, her call to participate "in real life" urges the viewer to move beyond the medium that currently connects them. Her on-screen appearance and its webcam aesthetic reflexively mirror the members who are watching her from their computers; as she sits and looks into her computer screen, the viewers sit and gaze into theirs. Even as it emphasizes the technological interface, the video also puts a human face on an organization that otherwise largely exists as a web page on the Internet. Like the house party, these member testimonials seek to personify MoveOn the organization. Their DIY aesthetic and their status as indexical moving images speak to a desire on MoveOn's part to reveal the people behind its organizing prowess—people who, like the spectator, can do it themselves. As Paul Arthur might claim, it is through these aesthetic limitations that the videos gain the "jargon of authenticity" they seek to impart.[65] Significantly, even an organization as rooted in digital technology as MoveOn still has consistent recourse to the power of the documentary testimonial to further its aims.

For Greenwald, *Outfoxed* represented the formalization of his move into documentary activism as the founder of a new production company and nonprofit political foundation, Brave New Films. Functioning until this point as Robert Greenwald Pictures, Brave New Films signaled that political documentary would take center stage as part of a larger, issue-driven nonprofit media company. Since then, the foundation has evolved in a direction remarkably similar to the collaborative model that Greenwald pioneered with MoveOn. Brave New Films focuses on media creation—from short, viral video series like the "Justice" series, targeted at specific issues around prison and justice reform, or the immigrant voices project, which captured short autobiographical stories in episodic form,

FIGURE 3.5. Elinor from *Host an Event!* In her testimonial video, she reflexively gestures around the screen to encourage the members she's addressing to click on the associated links of the webpage in which her video is embedded.

to longer, feature-length projects that tackle a similar range of issues. Some, like the "Rethink Afghanistan" project, started as a series of video installments but were later built into films. For each project, the organization partners with allied organizations working in the areas the films address, thereby building a support and funding network and often providing a screening venue or built-in audience network. The films are also targeted toward educational frameworks like classrooms, offering free screenings and course materials to supplement the information shown in a film. Beyond this, the foundation's website enlists users, much the way MoveOn does, to host screenings and function as information sources within their communities on the issues addressed in any given campaign. Nearly two decades after it was founded, the organization continues to grow. Its 2017 annual report details dozens of projects, many of which draw audiences and views in the millions. While the organization's focus is on media production and distribution, there is also a clear effort to integrate the organizational components into the process in a fashion similar to that of MoveOn.

Though similar, the two organizations do not necessarily compete with one another. Rather, they view themselves as partners in an ongoing struggle to create a more effective democracy. MoveOn positions itself to organize members to make their voices heard, utilizing old and new media in the effort to do so. Brave New Films seeks to empower independent media organizations to hold politicians accountable and make the voices of the electorate heard by bypassing the "traditional gatekeepers," as Greenwald himself put it. The irony in their evolution is that MoveOn's roots in new media technology have increasingly led it to forms of old-media moving images, and Greenwald's skills in old-media production have increasingly found an outlet on new-media platforms. The question remains: Does either form of media, old or new, actually achieve what both groups seek—political change?

DOCUMENTARY AGENCY AND
READ/WRITE CULTURE

Documentary film scholars and practitioners have frequently struggled with answering the exact question posed above. In what has become a classic essay in the field of documentary film scholarship, Jane Gaines posed a question that had hung over the heads of filmmakers and scholars for much of the existence of the genre: have documentary films changed the world, and if so, how? Referencing a 1995 study by Kirwan Cox for the National Film Board of Canada, Gaines notes that the "forty-eight scholars and filmmakers polled had difficulty thinking of any films that had actually 'changed'" anything, opting instead to point to films that had achieved some level of local influence. Undeterred, Gaines decided to conduct her own poll of scholars and filmmakers, but her results were largely the same. In lieu of films that had created change, they opted to discuss films that *should* have changed the world. She writes: "[A]lthough they could list films which had moved them personally, they could not be certain that these films had actually changed anything for anyone."[66]

Given documentary's long history of social activism, from the work of John Grierson on down, this result is surprising. As Gaines notes, it flies in the face of the desire of legions of filmmakers and activists who look to film as a tool for social transformation and social justice. While it seems absurd to believe that every film documenting a pressing social issue will produce a desired change, it seems impossible that eighty-plus years of documentary output had failed to produce a single, exemplary case study of a film that had.

Rather than a failure of the documentary form to produce change, Gaines's essay demonstrates not only the difficulty of the precise, quantitative cause-and-effect measurements of "media effects" more broadly, but also the emergence of a widespread skepticism regarding documentary film's relationship to notions of objectivity, truth, and reality. Indeed, as many of the essays in Gaines and Michael Renov's *Collecting Visible Evidence* anthology demonstrate, the entire modern period of documentary film scholarship—initiated, among other things, by the 1992 publication of Bill Nichols's landmark *Representing Reality*—is characterized by the assumption that the necessary relationship between documentary and reality is at best tenuous and at worst a fiction.[67] The scholars polled by Gaines had trouble finding a film that changed the world because the solid, indexical links between a film and the world at that particular poststructural, postmodern moment were under broader scrutiny. For those questioning the solidity of documentary's "truth claims," the ability of truth to be captured and then convince an audience was equally suspect. And, as the work of Errol Morris discussed in the last chapter exemplified, this skepticism about the relationship between truth and documentary is held not just by scholars but by many film-makers as well.

Beyond academia, however, a belief in and desire for documentary agency persists. In October of 2011, for example, an article in *The Guardian* subtitled "Can Documentaries Change the World?" provocatively echoed Gaines's question.[68] The occasion for the article was the announcement of the creation of the Creative Impact Awards by the BRITDOC Foundation, which aimed to "honor the documentary film creating the most significant impact in the world."[69] In an indication of the somewhat straightforward way the issue would be judged, BRITDOC (which renamed itself the Doc Society in 2017) elected a jury that included mainstream directors such as Morgan Spurlock, whose 2004 film *Super Size Me* claimed responsibility for forcing the restaurant McDonald's to remove its "supersize" option from its menu. As a marker of public perception regarding the agency of documentary film, the article set a fairly stringent standard for what it defined as cause and effect. Along with the example of Spurlock's film, Errol Morris's film *The Thin Blue Line* (1988) is also offered as an example of a film that was able to achieve direct results—in this case, the release of Randall Dale Adams from prison for murder.

Needless to say, this sort of one-to-one relationship between cause and effect in a given documentary film and its call to action is fairly rare. Many of the problems that documentary films seek to solve or address cannot be simply fixed with a single decision or outcome on the part of the government or a company. Films that have since won a Creative Impact Award, such as Eugene Jarecki's *The House I Live In* (2014) and Laura Poitras's *Citizenfour* (2016), present vastly complex problems for which no easy solution emerges. Even the two films the article cites as clear evidence of documentary agency present broader agendas than the release of a single individual or even an admittedly large product change by a major corporation. Such intangible, amorphous results are open to interpretation and refutation, and they certainly don't make for a clear and compelling newspaper article. But they are nonetheless the stakes most feature-length social-issue documentary films seek to achieve. The article gestures in this direction, speculating that perhaps filmmakers aren't so much interested in changing the world as they are in changing the minds of the people in it. Thus the question "Can documentary films change the world?" is at once too simple and too complex to answer. The mere act of bringing a film into the world changes it, even if only imperceptibly, and it would be difficult to argue that spectators remain entirely unchanged after watching a film, even if their opinions on a given issue are. *How* documentary films change the world is where things get complicated.

SOCIAL MEDIA FOR SOCIAL CHANGE

About the same time that BRITDOC was inaugurating its Impact Awards and *The Guardian* was pondering the possibility of documentary agency, scholars and journalists began asking similar questions of another new form of media: social media. In the October 4, 2010, issue of the *New Yorker*, a debate that had been quietly rumbling inside of academia for several years spilled out into the conversation of the general public. Malcolm Gladwell, the *New Yorker's* prominent and popular debunker of conventional wisdom and social statistics, argued that social media like Twitter and Facebook were incapable of producing meaningful social change or challenging the status quo. Using a history of the civil rights movement in the 1960s, particularly the lunch counter sit-ins that began in Greensboro, North Carolina, Gladwell claimed that true social change requires the presence of strong-tie relationships between social actors. Social media, built on weak-tie connections between disparate people, are consequently incapable of producing the level of commitment necessary to foster something like the civil rights movement. Gladwell acknowledged that weak-tie connections are capable of initiating lower levels of commitment from a greater number of people but maintained that such groups would never change the world. As Gladwell put it, "The Internet lets us exploit the power of these kinds of distant connections with marvelous efficiency. It's terrific at the diffusion of innovation, interdisciplinary collaboration, seamlessly matching up buyers and sellers, and the logistical functions of the dating world. But weak ties seldom lead to high-risk activism."[70]

Gladwell's argument arrived against the backdrop of the initial stages of what eventually would be referred to by Western media as the Arab Spring, which, among other changes, provided a less-than-clear test case for competing theories about the efficacy of social media in producing social change. On one side, the popular uprisings across the Middle East, reported in many major news outlets to have been organized and carried out using popular, widely available social media networks, seemed to confirm the predictions of scholars like Clay Shirky and Yochai Benkler, or journalists like Tim O'Reilly and Andrew Sullivan. Long predicted to change the world, events like the "Twitter revolutions" in Moldova and Iran seemed to fulfill technology's promise. On the other, less optimistic side, scholars like Evgeny Morozov and Golnaz Esfandiari pointed out that the role of technology in these events was misunderstood, and more a product of Western journalists' imaginations than activists on the ground. Gladwell clearly sides with the more dystopian outlook, even going so far as to maintain that substantive social change can occur only as the product of closely integrated hierarchical organizations that are run in a rigid, top-down fashion. While a decentralized, crowdsourced network might produce a project like Wikipedia, Gladwell concludes, "[t]he things that [Martin Luther] King needed in Birmingham—discipline and strategy—were things that online social media cannot provide."[71]

The exact role of social media in the Arab Spring and in other populist demonstrations, from the Occupy movement (2011–12) to Black Lives Matter (2013–), continues to be a topic of debate among scholars and activists.[72] Most acknowledge that the technology plays a definite role, but also reserve credit for people and offline forms of action like protests and petitions.[73] Theses debates strongly echo the fears that seemed to plague Gaines in relation to documentary film. No one disputes that Facebook has changed the world; the question comes down to the extent to which it upholds, or disrupts, the status quo in a given political context or on a specific issue. For Shirky, the major impact of technology on society lies in allowing people to organize differently, and these new types of organization can inevitably lead to social change. Shirky's *Here Comes Everybody* collects countless examples of groups coming together in an ad hoc fashion to establish new resources and institutions (like Wikipedia) and disrupt the status quo for others (like newspapers). In facilitating connections between individuals, new communications technologies enable novel collaboration and social organization. He claims, "More people can communicate more things to more people than ever before, and the size and speed of this increase makes the change unprecedented."[74] Given the breathless enthusiasm with which much of the book is written, it's easy to see why Shirky is often accused of utopian technological determinism. And yet, Shirky's pronouncements are often more circumscribed than his critics acknowledge. Gladwell, for example, faults him for celebrating the positive potential of technology in recovering a lost cell phone as though this somehow portended a revolution but skips over his discussion of the Catholic sex-abuse scandal or the Howard Dean campaign—both cases of legitimate social change, if not on the scale of the civil rights movement.

Even as it potentially empowers individuals, however, the emergence of social media has also placed more power in the hands of the state. For Evgeny Morozov, the Internet has evolved over the last two decades toward greater and greater levels of state administration over the policies and potential that online communication has to offer. The illusion of the democratizing, connective power of technology (what he calls "cyber-utopianism") leads to misguided policy and regulation of the Internet (in his terms, "Internet centrism").[75] The abstract belief that the Internet and the cohesive, spontaneous networks and forms of expression that it fosters are unequivocally conducive to democratizing principles is what fools individuals into complacency about the policies and principles that should be used to regulate it. This opens citizens in repressive states to greater and greater levels of government scrutiny and blinds those in the West to the surveillance and censorship taking place in their own backyards. The typical dichotomy drawn between the relatively democratic West and more authoritarian countries like Iran, Myanmar, and China hides the fact that competing forces are at work—forces that in both environments are working toward greater control, censorship, and surveillance of the individual by the state. So long as a country like Moldova allows for a certain level of free expression, not only will it identify problematic individuals for closer monitoring, but it will also give the rest of the population the illusion that they enjoy complete freedom online.[76]

Moreover, fears of the commercially disruptive nature of open networks result in greater censorship and scrutiny by the government. Consider the debate in the United States over the Stop Online Piracy Act (SOPA), and the Preventing Real Online Threats to Economic Creativity and Theft of Intellectual Property Act (PIPA), both of which were targeted at online file sharing but were criticized as unnecessary government intrusions into the privacy of individuals. Hence, the West is looking more and more like China even as China is looking more and more like the West. Morozov writes: "Anyone designing [online regulations] should be aware of some major inconsistencies between the strong anti-regulation impetus of Western foreign policy and the equally strong pro-regulation impetus of Western domestic policy."[77] Privacy and anonymity are the expectations imposed on US companies like Google and Apple operating in domains as diverse as the European Union and China. But many media companies, keen to protect their intellectual property from anonymous file-sharing sites, work to oppose these same principles within the United States itself. Given the number of variables (business model, political context, policy), generalizations remain elusive.

These commercial contradictions also connect with a common inconsistency in discussions about the ability of the Internet to empower individuals and disrupt existing organizations. As the subtitle of Shirky's book *Here Comes Everybody* asserts, the struggles that we see taking place in networked environments are the result of "the power of organizing without organizations."[78] This newfound power puts the existing organization, be it a business, industry, or government institution, at odds with a newly united group of previously disparate individuals. The work of scholars like Shirky and Lawrence Lessig are filled with examples of this newfound power at work. But, as Morozov repeatedly stresses, context matters. What works in one political and cultural context may not translate across borders. And what's true in the commercial sector may not apply to government institutions. Too often in cyber-utopian discourse, these contextual details are ignored in favor of a string of success stories pointing to the seemingly limitless potential for the Internet to change the world. This leads commercially disruptive trends to be conflated with a politically disruptive form of individual agency that may or may not exist in any given political context. The effect that blogs and online file sharing have had on newspapers and record companies is undeniable, but the impact that Twitter and Facebook have on emerging and established democratic societies is more elusive.

Too often proponents or critics of the "democratizing" power and potential of new technology frame the discussion as though the entire future of democracy itself would be determined (or not determined) by the emergence of a particu-

lar technology. The idealist, revolutionary ethos behind the open-software movement and early networking technology seems to have infused the debate around
technology's impact with an all-or-nothing tone.[79] Both stances ignore the obvious middle ground, where existing social forces shape new technologies, and vice
versa. Gladwell demonstrates both extreme stances on the issue when he states,
"Activists used to be defined by their causes. Now they're defined by their tools."[80]
As he points out, the church gathered people and disseminated information just
as Twitter or Facebook would today, and yet we don't call what it did the "church
revolution."[81] But this doesn't mean the church was incidental to the advent of the
civil rights movement. Calling an event the "Twitter revolution" is surely absurd;
denying that Twitter played any role, equally so.

THEORY MEETS PRAXIS

These universal pronouncements about the power of social media and other
technologies to cause (or not cause) social change demonstrate the need to
consider such questions on a case-by-case basis. Examining scenarios that put
newer technologies alongside older ones allows us to examine the historical
and cultural environment in which these two forms collide. This is where the
partnership between MoveOn and Greenwald becomes essential to understanding the way two particular forms of media technology mutually influence one
another. Three innovations in particular emerge from their collaboration and
seem to have shaped the current hybrid approach that both organizations have
taken. Each of these strategies implicitly capitalizes on many of the strengths
that scholars have identified within these tools and mitigates the potential pitfalls that Morozov and others identify in relation to utilizing social media to
achieve social change.

The first outcome of the collaboration that both organizations exhibit is a clear,
early mastery of many of the elements of what later came to be referred to as participatory culture. MoveOn in particular demonstrated clear, early innovation in
several key areas long before other commercial websites developed and perfected
them as tools for building an audience and monetizing its attention. The organization was operating as a type of social network before MySpace and Friendster
developed or Facebook perfected any particular model of community. MoveOn's
genesis into a progressive powerhouse coincides almost perfectly with the early
period that José Van Dijck identifies as the genesis of social media platforms (circa
2001).[82] Van Dijck's working definition of the social media platform as a sociotechnical configuration that "connects people to ideas to things to money" reads
almost like a description of MoveOn's attempts to use digital tools to foster what it
refers to as "people-powered democracy": fund-raising, petitions, demonstrations,
media sharing—all focused within the framework of progressive political organizing and network building.[83]

This description is broad enough to fit many different organizations, of course, but MoveOn's initial direction and consistent focus are enough to mark the organization as an early mover in this space. In its roots as an e-mail petition, MoveOn seems to have understood the strength of leveraging people's existing relationships. This leverage allowed the group to grow at viral speeds, and the general shift of politics in the United States after the Monica Lewinsky scandal gave members something to focus their energies on.

The second innovation, the adoption of the "house party" model, also worked as a form of location-based organizing similar to what other services like Meet-Up would eventually develop. And finally, in its model of funding—and, in the case of the *Outfoxed* production, creating—media items and political campaigns, MoveOn developed early models of crowdsourcing and crowdfunding before other services like KickStarter and Mechanical Turk emerged.

All of these technologies (social networking, location-based media, and crowdsourcing) became essential features of what we associate with Web 2.0 and participatory culture more broadly, but all three, under different names, are clear features of grassroots political organizing. MoveOn's achievement was to take the means and methods of grassroots organizing and adapt them to emerging technologies more fully and successfully than anyone else had up to that point.

For its part, Brave New Films anticipated participatory culture by evolving a collaborative, responsive model of filmmaking that I will refer to as a form of remix documentary. Its output demonstrates a clear willingness to adapt its production and distribution methods to work within the evolving domains of political organizing and participatory culture. Its responsiveness to emerging political issues seems to answer the call Jane Gaines put forth in her provocatively titled article "The Production of Outrage" in 2007. Here she argues that the production of a film is itself is a form of social action.[84] She outlines the need for filmmakers to "image out," in the same way that people might speak out against a given social atrocity, by creating films that address the need and the problems inherent in a set of historical events. Images of suffering and other atrocities, recontextualized through the documentary film, have the power to initiate and inspire social change that they might not in other contexts. In this way, film is able to "use the world to change the world."[85] Rather than simply being a middle step that might advocate for social action, film itself is a form of social action.

Gaines was specifically responding to what she saw as a level of social fatigue with images of the war in Iraq, as a form of iconophobia, or as a war on images of war. But her call for the further "production of outrage" perfectly describes the particular call to action that Greenwald felt when he made the transition to political filmmaking after the 2000 election. The emergent iconophobia in 2007 was a result of the massive quantity of digital imagery now available, a quantity that inures spectators to the problem instead of inspiring political action.[86] But rather than destroying or limiting exposure to such images, many of the Brave New Films projects repurpose such images to break through the noise and lay bare the social contradictions they conceal. This method of reformulation or recontextualization sits at the heart of several of Brave New Films projects, including *Outfoxed* and *Rethink Afghanistan*.

This method of recontextualization eschews a more traditional, representational conception of documentary (film as a "mirror" or "window" of the world) in favor of something much closer to Lawrence Lessig's description of remix. Just as remix depends upon preexisting material and disparate fragments that are pulled together in novel formulations, documentary film must utilize and recontextualize preexisting representational tropes and narratives to articulate novel arguments about the world. A documentary about the Iraq War, for example, wouldn't necessarily reveal something previously unseen but would, rather, try to break through the existing representations of the war by posing an alternative formulation and inviting a response from the audience. Indeed, Lessig's description of remix, or what he calls "Read/Write" culture, could be mistaken for Gaines's descriptions of documentary culture: "RW culture extends itself differently. It touches social life differently. It gives the audience something more. Or better, it asks something more of the audience. It is offered as a draft. It invites a response. In a culture in which it is common, its citizens develop a kind of knowledge that empowers as much as it informs or entertains."[87] While the concept of remix has been heavily associated with a certain aesthetic common to the sort of "mash-up" videos one finds commonly going viral on sites like YouTube and Facebook, at a more fundamental level it is rooted in a desire to challenge the existing cultural and political narrative through recontextualization.

In addition to channeling the remix zeitgeist in its film work, Brave New Films was also reworking other elements of the industry. While still producing feature-length films that formally fit within the boundaries of documentary practice, its distribution methods rapidly evolved over a few short years to disintermediate what Greenwald called the "traditional gatekeepers" at film festivals and studios. His means were largely the organization that MoveOn was simultaneously building and the new uses of technology that it was adapting. At the center of these innovations, particularly the house party and crowdsourced production, were Greenwald's films. Thus, politics was the ground that enabled the coevolution of documentary film and technology for both organizations.

The second exceptional feature about the Greenwald–MoveOn collaboration is its utilization of what we might refer to as a coalition model of documentary production. As David Whiteman points out, within a social-movement organization, the documentary form often forces stakeholders to synthesize and articulate their issues and concerns and to work to establish a common blueprint for their competing desires and outcomes.[88] According to Whiteman, each stage of documentary production, distribution, and exhibition might have the effect of bringing together different parties related to a particular issue and inviting them to engage in dialogue. Making a film, it turns out, can educate filmmakers and participants in the same way that seeing one can educate audiences. Whiteman calls this the "coalition model" of documentary film, a model that describes the social fabric the film can weave among disparate or previously disconnected social groups.[89] Like the broader coalition of nations that would eventually invade Iraq twice in 1991 and again in 2004, Whiteman's coalitions are often made up of disparate, heterogeneous groups connected to a broader social issue like labor relations in the southeastern United States or strip-mining in rural Appalachia. Once in conversation, Whiteman demonstrates, such groups often remain united in their common cause. Indeed, MoveOn and Brave New Films offer a clear example of the kind of coalition building that Whiteman describes. Rather than changing people's minds, such documentaries may have their greatest impact on those who already agree with the film. While such films might be "preaching to the choir" of sympathetic audiences, their overall aim in Whiteman's model is to unite the choir and to get it to sing louder.

The form of remix documentary that Brave new Films and MoveOn produced relies on bringing disparate, preexisting elements together into novel configurations and relations in much the way that remix or Read/Write culture does. Whether "elements" here refers to the social actors involved in the production/distribution/ exhibition of the film or the disparate images and media fragments that make up the text itself, documentary as remix circumvents the traditional identification of photographic indexicality as an essential component of documentary truth. This more experiential, process-based form of documentary agency acknowledges and even depends upon the presence of alternative representations of the world for its own intervention. Rather than relying on a privileged connection to the world for its form of truth, a film like *Outfoxed* seeks a version of truth about the falsity of a network like Fox and the untrue images it broadcasts.

The type of documentary remix that Brave New Films and MoveOn put forth nonetheless maintains documentary's traditional identity as an alternative form of media expression that circumvents and subverts the traditional, commercial monopolies that dominate the media landscape. This model of documentary as "user-generated content" has existed at least as long as the Workers Film and Photo League first emerged in the 1930s. As Lessig claims, Read/Write culture is neither new nor isolated to any particular form of technological media. The impulse to "speak back" to a dominant cultural text is already common to language and writing. For Lessig, all that has changed is the ability for the average person to access and alter forms of modern media traditionally protected by significant technological barriers. Less emphasized by Lessig is the extent to which remix culture might be capable of producing novel forms of social organization and consequently social action. This, however, was the focus of the type of documentary remix that Brave New Films and MoveOn produced throughout their collaborations with one another.

It is important to note, however, that this remix takes place as a result of the two organizations coming together and collaborating with one another. Put differently, it takes both the presence of Greenwald's documentaries and MoveOn's digital organizing to achieve this. None of Greenwald's films during this time contained radically novel material or arguments that no one had seen, nor did they contain any investigative "smoking guns." They were effective because they put into a single text what many people, MoveOn members in particular, already suspected: that Bush stole the election, that the Iraq War was unjust, that Fox News was biased. They served as rallying points for MoveOn members to come together and mobilize in opposition. The "remix" of ideas and groups depended on both the films and the virtual organization in which they were seen and produced.

Finally, the MoveOn–Greenwald collaboration also demonstrates a solid middle-ground case between the all-or-nothing extremes of the debate over technology and social change. Much of MoveOn's activity seems to fit solidly within the model of "clicktivism" that critics of technologically fueled social activism decry. That is, that signing petitions, forwarding links, donating small amounts of money all have the appearance of political participation but don't necessarily produce the types of momentous social change that critics like Andrew Keen or Malcolm Gladwell describe. And yet, within and alongside these activities, and very much dependent upon them, is another set of activities that fit very much within the model of traditional, strong-tie activism. This is where the house-party model once again becomes essential. The "nationwide" screenings of Greenwald's films provided MoveOn members with an event that let them take their virtual connections and map them onto a series of local, geographic areas. Since these house parties formed the basis for other offline activities (protest marches, phone calling campaigns, etc.), the event of the film screening acted as a conduit between the online and offline worlds in a way that enabled MoveOn to translate the size and scale of a network of weak-tie connections into the commitment and motivation that comes with strong-tie, face-to-face interaction.

While the MoveOn–Brave New Films collaboration might not have changed the world as radically as either might have hoped, it nonetheless advanced to some degree the agenda of the progressive causes for which both advocated. And it certainly advanced the organizational structure and approach that both adopted in utilizing media as a form of connective thread between members, audiences, and their political leaders. As these forms of organizing spread across the political spectrum in the United States and other countries, this proved to be a lasting change indeed.

4

"States of Exception"

The Paradox of Virtual Documentary Representation

Shortly after his inauguration in January of 2009, President Barack Obama made headlines by signing an executive order that pledged to close the prison camp at Guantánamo Bay, Cuba, and revise US policies for questioning and detaining terror suspects. These moves sought to signal his administration's shift in waging the "war on terror" declared by his predecessor. He stated at the time: "The message that we are sending around the world is that the United States intends to prosecute the ongoing struggle against violence and terrorism" but will do so "in a manner consistent with our values and our ideals." In language intended to signify his rejection of the Bush-era binaries that drew stark contrasts between opposing sides, Obama added, "We continue to reject the false choice between our safety and our ideals."[1] After several attempts at closing the base failed within that time frame due to concerns over where to move prisoners, the administration admitted that finding a solution would take longer than expected and would contain provisions for extrajudicial trial and "indefinite detention." In spite of what may have been a sincere and honest attempt to undo the ethical and political damage created by Guantánamo, once established, such a place proves rather unyielding to the changing political tides that surround it.[2]

In the early months of his second term in office, long after the issue of Guantánamo had moved to the political and national back burner, Obama once again faced questions relating to his administration's policies on trying and prosecuting perceived enemies in the war on terror. This time, however, the issue at stake was the use of unmanned aerial vehicles to track and kill enemies of the United States without trial or official oversight. Rather than detain and imprison terror suspects, it seemed that the so-called kill policy simply eliminated them outright. Attempting to draw attention to the issue, Senator Rand Paul took the extraordinary step of filibustering the Senate for thirteen hours, questioning the legality of using drones to kill US citizens on American soil without any congressional or judicial oversight.[3] Citing the need for national security and the authority to act quickly, the Obama administration, in its defense of its policy, sounded reminiscent of many of the rationalizations of its predecessor's. Many began to wonder if the "regime change" of 2008 had delivered much change at all.

The legal and ethical complexity of both the Guantánamo prison camp and the drone policy demonstrate the bizarre political and legal limbo entailed in waging the ongoing war on terror. Faced with a nonconventional enemy, one free of state identification and capable of blending in with existing populations, the US government eliminated any pretense of upholding long-standing ethical and legal norms around individual privacy, state transparency, and international human rights and political sovereignty. Starting with the passage of the Patriot Act, and extending through an ad hoc series of memos, policy declarations, and public and covert actions, the government began tracking, detaining, torturing, and killing those it suspected of further terrorist acts against the United States. In the process of justifying and carrying out these actions, it created a patchwork assemblage of legal and logistical anomalies—"vanishing points" that enabled the US government to execute and expand its war on terror.[4] Outside of the legal and ethical questions they present, drones and other forms of weaponized technology further sit at the heart of an increasingly technological arsenal that utilizes video-game and virtual technologies to recruit, train, and equip the soldiers fighting in and across the various "battlefields" that make up the war on terror.[5]

For those opposing the war and the way it is/was being fought, these same tools and technologies offer a means of exposing and opposing these policies. This chapter will contrast two radically opposed approaches that use virtual technology to simulate, document, and engage the bizarre battlefields of the war on terror: the US military's integration of networked technology and virtual environments as exemplified by the *America's Army* video game and its expansive drone program, and the *Gone Gitmo* project created on the Second Life platform. *Gone Gitmo* was an early attempt to use the sandbox of a virtual world to expose one of the vanishing points in the military's arsenal of off-scene spaces: the Guantánamo prison complex, which had become ground zero for the government's unprecedented and unlawful imprisonment of terror suspects after 9/11. Created on the virtual platform Second Life, *Gone Gitmo* demonstrates the problematic power of pushing documentary reality into a space of fantasy and play, challenging the easy distinctions between real and virtual spaces. *America's Army* and its drone program travel these same boundaries. Shortly after 9/11, the military utilized the extensive assets it had invested in training simulators and virtual combat technologies over the previous decade to launch a free, multiplayer first-person-shooter video game to the public. Dubbed *America's Army*, the game was a clever tool for training and recruiting a new generation of soldiers to fight the war on terror.[6] Even as the game achieved these aims with a blockbuster degree of success, its prominence and purpose also attracted the attention of artists and activists seeking to complicate and critique the easy exchange it afforded between real and virtual conflict—a relationship further complicated by the military's increased reliance on drone technology to project power even farther from the soldiers tasked with fighting these battles.

Though the landscapes of the war on terror have become inaccessible to the type of optical recording technology traditionally used to wage and oppose war, both of these projects attempt to relay players back to reality in a way consistent with traditional documentary film. And yet, the military's massive recruiting and training efforts in *America's Army* end up distorting the reality of warfare, whereas the activist exposé *Gone Gitmo* skillfully plays on the realities of virtual representation to critique the military's policy of torture and indefinite detention. This difference enables one to offer a critique of the other, and together both demonstrate the possibility of maintaining a documentary impulse in the absence of the traditional documentary image.

TRAVELING FROM GUANTÁNAMO TO *GONE GITMO*

The question that Obama faced when he took office—how to handle Guantánamo—was also one that faced the myriad of political activists who opposed Guantánamo's existence, a group that included civil rights attorneys and human rights groups as well as journalists and documentary filmmakers. The military had repeatedly blocked requests for media access to the base and adequate legal representation for the men imprisoned there. This left such groups struggling to find a way, legally and visually, to represent Guantánamo in order to draw public attention to the issue and the individuals involved. Some of the unique tactics these groups utilized to "represent" Guantánamo offer insight into the complicated political issues surrounding it.[7] One such solution—the *Gone Gitmo* project on the virtual platform Second Life—demonstrates that the nature of such environments uniquely mirrors the paradoxical political nature of the physical place that they re-create.

Gone Gitmo, the product of a collaboration between University of Southern California (USC) graduate student Nonny de la Peña and USC visiting professor Peggy Weil, was created in 2007 during a residency of theirs at the Bay Area Video Coalition.[8] The guiding idea behind the project was to allow users of the Second Life platform to experience the Guantánamo Bay prison camp "firsthand" by virtually re-creating the prison in exacting detail. As the pair explained at an American Civil Liberties Union (ACLU) event featuring the work:

Our purpose is to raise awareness, initiate discussion and educate on habeas corpus issues by making a virtual but accessible Guantánamo Bay Prison in contrast to the real, but inaccessible, U.S. prison camp. We are using Second Life to expose a substantially new audience to these issues by extending the methods and images from documentary filmmaking into new online, participatory environments.... Our overriding philosophical challenge is to communicate a gravely serious matter in a medium known for games and entertainment. . . . As artists, we confront how to portray the practices in Guantánamo effectively and design an experience that does not trivialize torture (we will not torture your avatar) but will provoke thought and insight into the complicated issues surrounding detainees' rights.[9]

When users of Second Life typed in the project's address, their on-screen character (or avatar) was immediately hooded and transported, with the screen darkened, to a holding cell in the prison simulation on the project's space. Once there, they were free to explore the virtual space, which included links to numerous articles about the prison as well as a video feed running testimony by the few detainees who had been allowed to speak on camera about their experiences in the real prison. In order to leave the prison, visitors could simply enter the address of another Second Life location, whereupon their avatar would "fly away" from the prison—a form of individual autonomy many of its real-life inhabitants did not and do not experience. This description and the statement given by the project's creators demonstrate that *Gone Gitmo* is something of a paradox, a hybrid space that blends different media together to access each term through its opposite: the real by way of the virtual, the inaccessible via the open, the gravely serious in the space of play. In short, it answers a paradox with a paradox.

Second Life presented a provocative opportunity for this type of project. When it first came online in 2002–3, the platform puzzled many people because it wasn't necessarily clear what one was supposed to "do" there. A cross between an open-world or sandbox-style game (where users are allowed to freely explore the virtual-game world rather than play through with a specific goal in mind) and a collaboratively constructed environment, the platform offers users an enormous degree of freedom to decide how they will spend their time and to discover what sorts of spaces and events it might contain. At the height of its popularity, it commanded an enormous amount of creative, critical, and commercial attention, attracting major brands, universities, and artists alike.[10] Many imagined it as a sort idealized utopia—a space one could inhabit without the limitations that time, space, and resources place on us in our offline worlds. A virtual prison, in other words, may have been within the bounds of what one *could* create, but it fell well outside the interpretation that many had for the blank canvas that the platform presented to users.

The documentary tendency at work in *Gone Gitmo* is immediately apparent in its attempt to point us toward the real, historical world—in opposition to the imagined, fictional world typically on offer in Hollywood cinema. Fundamentally, we

FIGURE 4.1. The screenshot depicts the *Gone Gitmo* space in Second Life. The other image shows the actual space. Nonny de la Peña and Peggy Weil relied on satellite imagery of the prison to re-create scale and layout, and rare press photographs like this one to fill in the detail about construction materials. Photo credit: Petty Officer 1st Class Shane T. McCoy, U.S. Navy.

think of this connection to reality as a product of film's ability to faithfully record reality (its cinematic indexicality) and documentary's visual representation of historical events. But other formal methodologies have also been used to achieve this same end, including reenactment, interview, testimony, and even animation. Moreover, many of the cinematic conventions that foreground the expressive con-structedness of the medium—like montage, subtitling, and frame composition— are components of documentary. Hence, the reality that appears in documentary arrives via many fictional routes. While it lacks the optical photographic indexi-cality of a film, *Gone Gitmo* seeks to represent reality as faithfully as possible. It re-creates the physical materiality of Guantánamo in precise detail, from the size and layout of the prison cells, to the details and building materials that are used in each of the spaces, to the orange jumpsuits worn by detainees.

Though one might convincingly argue that certain historical epics seek this same degree of material precision in order to open onto the most fictional of fantasies (James Cameron's *Titanic* [1997] comes stubbornly to mind), comparing *Gone Gitmo* with the rest of Second Life throws the distinction into greater relief. If we take the proponents of Second Life's ability to fulfill one's physical, material, and sexual desires and fantasies at their word, then many of its locations seem to mirror the purpose of mainstream fiction film. This starkly contrasts with a space like *Gone Gitmo*. One allows users to escape or alter reality; the other seeks to remind them of it.[11] If Second Life is a Hollywood-esque space of fantasy and play, *Gone Gitmo* is its "discourse of sobriety."[12]

Gone Gitmo further manifests a documentary impulse or tendency through its attempt to intervene in issues of social justice. To be sure, plenty of documentary films speak to issues other than those in the political arena, but documentary's attempt to inform, persuade, and advocate on this front certainly represents one of the genre's major categories.[13] Like a traditional political documentary, *Gone Gitmo* offers us both a specific political position (that Guantánamo should be closed) and a defined call to action (with links to write one's senator and specific interest groups to support.)[14] Beyond its connection to the real and its political sympathies, *Gone Gitmo* further shares several formal similarities with docu-mentary form. The first is its ability to pull in multiple forms of media in order to marshal its argument and achieve its aims. Similar to a documentary that utilizes archival footage, interviews, newspapers, still photographs, sound recordings, and reenactment, *Gone Gitmo* contains elements of all of these things, includ-ing poetry written by detainees, newspaper headlines (with links to the stories), fragments of footage from traditional documentary film (taken from Nonny de la Peña's 2004 *Unconstitutional*), and reenactment (the hooding of the avatar and its placement in a C-17 transport plane when teleporting to the *Gone Gitmo* site). While the question of perspective, point-of-view, and omniscience in relation to virtual environments is complex, suffice it to say that Second Life contains its own version of camera angle, depth-of-field, and so on—all roughly analogous to the role these qualities play in film language.[15]

But if *Gone Gitmo* exhibits these similarities to documentary film, there are also a number of points where it expands the limits of documentary. The first is the project's ability to stay current. As mentioned, the project includes a number of places where news feeds and other media sources are pulled in, similar to a Ken Burns–style pan and scan of these materials in a documentary. Unlike a film, however, which remains tied to a given historical point upon its completion, *Gone Gitmo* was both updatable and auto-updating. News feeds enabled visitors to be up-to-the-minute on the information they receive. The space itself was constructed and reconstructed several times to reflect changes in the physical layout of the Guantánamo camp itself, lending its representation a temporal mutability that would be impossible in film.[16] This gives it a sense of temporal currency that mimics the sense of time on the Internet more broadly (a quality that also has drawbacks that I'll discuss below).

The second advantage that *Gone Gitmo* offers over traditional film lies in its predominantly nonindexical form of representation. In spite of the best efforts of Linden Labs to make Second Life as photorealistic as possible, one would never confuse it with a photographic representation, much less reality itself.[17] On one hand, this extreme mediation removes us from the real that documentary always seeks, and yet it also reminds us of the limits of the experience we are given. We never confuse the representation with reality—a clarity that keeps visitors from overindulging in a potentially delusional empathy for the victims and focuses debate on the issue itself. One might come away feeling as though one better understood the issues involved, but it seems unlikely that anyone would feel as though he or she had experienced what the detainees in Guantánamo have experienced. In discussion boards populated by people who have been to the site, it is amazing to find very little of the cynicism regarding source material and omission that seems to haunt a filmmaker like Michael Moore.[18] Instead of debating the accuracy of the representation, participants turned to the ethics or efficacy of Guantánamo itself. Users seem to have no problem making the jump from the project itself to the issue involved, despite the mediated form the representation takes. While this doesn't necessarily promote a more civilized debate (something message boards rarely seem to achieve), it does enable the discussion to avoid getting mired in the form of the representation and instead engage the substance of the issues that it is raising.

The third advantage of *Gone Gitmo* over a traditional documentary film is the spatial access that it offers its users to the space of the camp. Under the veil of security, Guantánamo has been notoriously off-limits to outside observers from the United Nations, human rights groups, and the media.[19] In the absence of the ability to record the location itself, virtual environments offer an excellent opportunity to open up otherwise invisible spaces. Furthermore, the user control and three-dimensional rendering offered in such an environment enables visitors to experience the project at their own pace, allowing them to linger or skip past different elements in a nonlinear, undirected order. Moreover, once one becomes used to moving around in Second Life, the particular sense of space it offers surpasses what we experience through two-dimensional images. An image of something like a prison cell can *look* small; trying to move an avatar inside of one begins to *feel* small. This clearly doesn't replicate the experience of actually being in the cell (with little to no hope of release), but neither does a film.[20]

As a form of representation, *Gone Gitmo* is thus a hybrid gesture that extends documentary's activist impulse into a medium that transcends some of its traditional limits. There are, however, some very clear limitations to working within Second Life as a platform, and in virtual environments more broadly. The primary drawback concerns the availability and accessibility of the project as compared to a film. Any user who wanted to experience *Gone Gitmo* as it was intended needs to have a fairly robust computer, a broadband Internet connection, and enough storage space and permission to download and install the Second Life application. After this, the user must register for an account (sharing personal data) and spend time learning the environment and its navigation tools. Weil and de la Peña were able to mitigate some of these factors in museum exhibitions and other venues by providing machines or projecting a recorded machinima of another user navigating the space.[21] While these workarounds opened the project to a wider audience, they obviously compromised something of the intended experience. Film and video certainly suffer from their own accessibility issues, but given their age, many of these issues have been addressed.

The availability of the project is, however, another matter. Since it was created on a commercial platform, the project had to be actively hosted in order to remain available to users. This required ongoing funding to cover the cost of the Second Life server space, whatever Linden Labs decided that cost would be. While the project received several development grants as well as donated space on different "islands" within Second Life (commercial accounts purchased by other groups), eventually the project ran out of options and disappeared off of the platform. Thankfully, a good deal of documentation remains due to the efforts of journalists like Draxtor Despres, but nonetheless the project no longer exists in its original format. This of course is a problem with any experimental technology, and certainly one that has confronted artists and curators working in digital forms like net art or CD-ROMs.[22] The interdependent framework of a website, dependent upon multiple layers of hardware and software from multiple commercial companies, has an abundance of failure points that can doom a project to obsolescence. This is compounded for projects dependent on a private, closed platform like Second Life, where the fate of the work is tied to the success or failure of the company that hosts it. These obsolescence and availability factors make digital art delivered through platforms to individual computers at once more available but also more fleeting than a traditional film text. Ironically, the virtual Guantánamo closed before the real one that it was protesting.

Considering the advantages and disadvantages of virtual platforms like Second Life, we might wonder why Weil and de la Peña chose it over a more traditional, established medium. After all, both had extensive experience in film, photography, and video, de la Peña as a producer and director of feature-length documentary and Weil in different gallery and installation projects. A more straightforward documentary would have been well within their grasp. So why choose Second Life? Notable in the artists' statement quoted earlier is the reassurance that they will not "torture your avatar," and in light of the forms of mobility discussed above, it is clear that they do not imprison or detain it, either. This is a striking absence, given that torture and unlawful detention were the primary purpose of the real Guantánamo and the key point of its political controversy. But the reassurance speaks to the potentially strong identification that many users have with their avatars. For many, Second Life is manifestly about exploring different identities and social positions other than their own—a possibility that was a prominent part of the discussion surrounding it in the height of its popularity. And this remains the case well into Second Life's second decade on the Internet.[23] For the avid enthusiasts on the platform, *Gone Gitmo* offered a way to understand something of the situation confronting the Guantánamo detainees.

We can further see the appropriateness of building a virtual Guantánamo by delving into the more perplexing aspects of the real Guantánamo—a place I earlier described as a legal and political paradox. Prior to 9/11, many people considered state-sanctioned torture and uncharged imprisonment legal, if not logical, impossibilities. And yet, as Giorgio Agamben has argued, such "states of exception" lie at the very foundation of political sovereignty in every state, including Western democracy.[24] Drawing on Carl Schmitt's formulation in *Political Theology,* Agamben argues that the "state of exception" is the political and legal framework, present in all democracies, whereby the leaders of the state can nullify the existing constitution by declaring a form of martial law. This allows them to selectively and capriciously apply existing law and consign any specific group of individuals to whatever legal designation they deem politically expedient. (His designation for this is the "force of law.")[25] Far from being a simple clause in the constitution that may be amended, it is this exception that enables the rule of law itself to exist. Hence the paradox. Much of what followed 9/11, from the Patriot Act to Guantánamo and its detainees, has provided a textbook example for Agamben of the manner in which sovereignty exists simultaneously both inside and outside the law.

Where Agamben's work takes on a specific significance for the kinds of virtual spaces that *Gone Gitmo* and the other projects utilize is through his concept of "bare life" and the particular biopolitical turn that he sees at the root of twentieth-century sovereignty. The "homo sacer" is the figure that emerges within those populations that sovereignty has excluded from the polis, but over whom it continues to exercise political power. Such people, reduced to a form of "bare life," vulnerable and exposed to violence and injury, exist within, and are subject to, this barest bodily materiality. Torture, imprisonment, and execution within an extrajuridical framework are all legitimated through this particular form of sovereign power, which reaches its zenith in spaces like Guantánamo, Abu Ghraib, and the other vanishing points of the war on terror. The bodily, material reduction to bare life of such spaces presents a provocative, and potentially problematic, counterpoint in the virtual worlds and disembodied spaces that both *Gone Gitmo* and *America's Army* utilize.

Much of Guantánamo's paradoxical state rests on extending the political/legal incongruity imposed on the bodies of individuals to the geographical spatiality Guantánamo occupies. Even its various labels ("prison camp," "detention facility," etc.) point to the indeterminacy of its exact nature. The space this defines has been further elaborated by geographers such as Derek Gregory, who draws out the particular colonial roots of both spaces like Guantánamo and the production of "homo sacer." For Gregory, such spaces are constituted and legitimated through the same tradition that justified settler colonialism and slavery. The "state of exception" (as well as the "space of exception" that it produces) is the flip side of the logic of Euro-American exceptionalism that surfaces when one of these sovereign powers decides to overstep or ignore the norms of international law that it would otherwise enforce on its neighbors. Rather than a kind of lawless black hole, Gregory maintains that these spaces are highly circumscribed and quite closely confined by the "ligatures between colonialism, violence and the law."[26]

It is tempting to dismiss the euphemistic labels applied to these spaces as further examples of the extreme limits to which political rhetoric was driven under the Bush administration. But as Judith Butler points out in *Precarious Life*, each of these terms is carefully crafted to perform significant political and legal legwork, stripping these individuals of not simply their rights but even their status as human beings.[27] The refusal to mourn the 9/11 attacks in a way that included a consideration of what caused them forced us to deny any consideration of the position of the Other, a refusal that opened the door for such future actions as denying detainees any claim to fundamental legal and human rights. Interestingly, Butler ties self/Other together in a way that locates responsibility and morality with both sides simultaneously, a move that places it in the same sort of liminal position that I'm claiming both Guantánamo and the *Gone Gitmo* project occupy. Part of shaping the discourse after 9/11 in a legal and media framework was deciding whose voice would be excluded from the conversation, an exclusion that *Gone Gitmo* and other activist representations seek to redress. The irony of the Bush administration's ability to place things rhetorically into simple either/or terms ("You're either with us or against us") is that its policies proliferated in places and populations that are neither/nor any of the established positions. As the case of the Uighurs demonstrated, every case in relation to Guantánamo offers an exception to the rule, and exceptions in turn make the rules themselves entirely untenable and meaningless.[28] In short, as Agamben says, the exception becomes the rule.

Beyond simply exemplifying the paradoxical nature of political sovereignty, Guantánamo's very existence embodies legal and logical contradictions of its own. Consider, for example, that the US military even occupies a base on one of the few remaining Communist countries in the world, one that it has virtually locked out of any diplomatic connections for much of the last half century. Even as the Obama administration began to reestablish diplomatic ties and normalize relations with Cuba after Fidel Castro (moves the Trump administration has since largely reversed), there was very little question that the base would remain in US hands, regardless of what its primary purpose was. While the arrangement predates the Cuban revolution in the 1950s and the political enmity the Castro regime brought with it, the original lease on the land was a product of what Larry Birns calls the "19th century gunboat diplomacy practiced by Washington" in the wake of the Spanish-American War. Even then debates swirled as to whether the base would be covered under US or Cuban constitutional law.[29] With the fall of the Soviet Union in 1991 and the transfer of the Panama Canal in 1999, the base seemed to have lost any strategic benefit to the United States and was looked at for closure several times during Defense Department budget cuts in the 1990s. And then came September 11, 2001.

The very same aspects of the base that made Guantánamo such a perplexing place before the war on terror in both legal and political terms made it an ideal place afterward. This irony was already obvious as early as 2003, when feminist scholar Amy Kaplan described Guantánamo in a *New York Times* op-ed piece as "a territory outside U.S sovereignty, held in perpetuity, where the U.S. military rules[.] Guantánamo is a chillingly appropriate place for the indefinite detention of unnamed enemies in a perpetual war against terror."[30] Kaplan later demonstrated that this same logic of uncertainty suffused the various US Supreme Court opinions that dealt with questions around the legal jurisdiction of Guantánamo due to its uncertain geopolitical location.[31] The only political entity with any claim to sovereignty over the base, the Cuban government, was the only one completely lacking the political or military resources to exercise it. The precedent for detention facilities there had been in place since the early 1990s, when the base was used to house both Haitian immigrants fleeing the fall of Jean-Bertrand Aristide's government in 1991 and the influx of Cuban refugees captured on the open sea between the United States and Cuba seeking asylum. The notorious Camp X-Ray, in fact, had been used during this time to house HIV-infected Haitian immigrants who were denied asylum in the United States. When the first detainees from the invasion of Afghanistan began arriving there in 2002, their designation as illegal enemy combatants rather than as prisoners of war consequently denied them their rights under the Geneva Convention. These detainees fell into the same legal limbo that the base itself existed under for the last century. In Guantánamo, the US military had essentially secured the perfect place to settle the equally legally dubious individuals who would be brought there.[32]

The dark genius of Guantánamo's creation lies in its nature as a hybrid space of sorts, one that exclusively fits none of the existing categories and hence one that can't be dealt with according to any of the established laws or guidelines. Such hybrid spaces—those that elude a clear classification and are therefore impervious to shifting political climates and overt political action—are only increasingly common in the "state of exception." It is this relative political uncertainty and physical inaccessibility that makes Guantánamo the perfect subject to represent with immersive virtual technology. If the strength of *Gone Gitmo* is that it responds to a paradox with a paradox, then to a large extent the uncanny uncertainty of the virtual world that exists in Second Life provides the appropriate analogue. Depending on the source, Second Life is either an experimentally (dis)embodied utopia or yesterday's next big thing. Either way, after exponential growth through mid-2007, the online world peaked at about one million regular visitors. In late 2011, the last period in which traffic was reported for the site, it had approximately one million repeat visitors who spent a total of 124 million hours a month collectively exploring it.[33] While the figure has declined since, the site still attracts a dedicated user base of about eight hundred thousand people every month.

Descriptions of Second Life tend to characterize the environment in two seemingly contradictory fashions. On one hand, its similarities with the real world are stressed: people do all of the things "there" that they do "here," from working and shopping to socializing and traveling. On the other hand, it is characterized as being nothing like real life: physical constraints such as gravity, hunger, fatigue, aging, and illness are all optional indulgences. In short, virtual environments like Second Life are paradoxically hybrid places. Similarly, discussions of the "experience" of Second Life are equally vexed. The philosopher Hubert Dreyfus points out that the fundamental deficit in platforms like Second Life is their lack of embodied finitude.[34] For Dreyfus, virtual environments predicated on a user consciously controlling the gestures, emotions, and reactions of an avatar treat users as mind-centered subjects capable of exchanging one container for another. Thus, they succumb to the fallacy of Cartesian mind-body dualism. The technological promise of a body without limits is precisely what prevents virtual environments like Second Life from delivering much of the physical, emotional, and social sensation that we draw from embodied experience in real life.[35]

Taking the opposite tack, many users of Second Life utilize the environment it provides to achieve physical experiences that they are prevented from encountering in their offline lives. The anthropologist Tom Boellstorff readily admits that virtual embodiment is concomitantly different from real embodiment, but he maintains that it nonetheless offers certain users experiences absent from real life.[36] Focusing on the sociality involved in group interaction, Boellstorff's study outlines several instances in which the users' ability to change their bodily appearance in Second Life allowed them to experience an identity different from their own based on the reactions of others. For Boellstorff, the ability to explore different facets of oneself in persona play offered users forms of fantasy, empathy, and self/other exploration denied to them in real life, occasionally with lasting results for the physically embodied person at the keyboard.[37] The extensive reporting done on Second Life by journalists (at its peak the site produced a staggering five hundred news stories a day) almost ritually compares the differences between the limitations of users and the experiences they play out through their avatars, including basic activities ranging from walking and pregnancy to extensive body modifications.[38] Even users who don't engage in these activities feel that their virtual bodies are truer to some aspect of their self-perceived selves than their physical bodies.

Neither fully embodied nor disembodied, virtual worlds can place users in a zone of indeterminacy that forecloses some experiences while enabling others. For the users of these worlds, this is not a drawback. It is precisely this neither/nor status that makes Second Life an intriguing medium through which to explore the politics of a place like Guantánamo—one that can also extend the limits of a medium like documentary film. While virtual environments sever ties with film in multiple ways, their remediation ensures that some cinematic aspects remain, including point of view, camera angle, depth of field, and so on.[39] Moreover, projects like *Gone Gitmo* demonstrate that a documentary impulse not only survives on such new media but also is essential to the impact they achieve. Several qualities of *Gone Gitmo* manifest what I am calling the documentary impulse, and in many of these ways the project even transcends the limits of documentary.

But beyond offering just some form of empathy and experience in relation to Guantánamo, *Gone Gitmo* is, I claim, perfect for the task, and this is because its complexities and contradictions replicate the peculiar political complexities and contradictions of the camp itself. Neither fully embodied nor disembodied, neither real nor fantasy, its users neither empowered agent nor passive spectator, *Gone Gitmo* utilizes this platform to translate the paradoxical limbo that Guantánamo inflicts upon its detainees. Returning to the issue that Obama faced throughout his presidency—what to do with Guantánamo—we unfortunately find that *Gone Gitmo* offers nothing in the way of a solution, nor perhaps will it be any better at convincing those on the opposing side of the merits of its case than the dozens of films that have been made on the issue. It offers no technological utopia. But even if Guantánamo were closed and bulldozed as Abu Ghraib was before it, the issues and victims it contains would simply migrate to new places like Bagram Air Base, the "Salt Pit," or any other of the "black sites" that exist away from the scrutiny of the public. When Gitmo itself is gone, the relevance of a project like *Gone Gitmo* will remain, even if we have to teleport to a new platform in order to see it.[40]

WAR GAMES

While activists like Weil and de la Peña were utilizing virtual platforms to expose the inherent injustice of US government policies around torture and detainment, the government began adopting these same technologies to make the detention of enemies in the war on terror an irrelevant issue. Almost from the moment of the terrorist attacks on September 11, 2001, the military began developing and utilizing video games and other virtual simulation technologies to recruit and train soldiers, and even more advanced robotic technologies to track and kill its enemies on the battlefield. Many of these were of course already in various stages of development within the military and its various research agencies before 9/11. But as with so many other programs and policies, the attacks provided an occasion to accelerate the integration of these tools into mainstream military practice.[41] If the complexity of battling a stateless terrorist enemy legitimated (in the minds of some) the creation of surreal legal and territorial spaces like Guantánamo, these places and policies found an intriguing counterpart in the virtual spaces and tools the military used in the process.

The military's new virtual weapons systems challenged easy distinctions between real and virtual, carving out a complex space of new nonfiction media in the process. As the *Gone Gitmo* project had done, these tools pushed the boundaries around embodiment and representation in new ways, creating challenging new ethical, physical, and experiential zones for the human bodies on these new battlefields. Where *Gone Gitmo* focused on using the complexity of virtual embodiment to approximate the physical and legal precarity of the people subject to US policies around detention and interrogation, both *America's Army* and its drone program used similar technologies to recruit and train the soldiers carrying out these policies. And, as their forerunners had done with Weil and de la Peña, these new hybrid, virtual spaces attracted the attention of activists and artists seeking to counter and call out the military's use of these same tools. In the course of modernizing its arsenal, the military created a set of paradoxical policies and experiences for the soldiers tasked with utilizing these new weapons, as well as a new terrain that would allow activists and artists to respond and intervene to disrupt these policies.

The entrance of this new breed of weapons into mainstream military practice arrived on a wave of popular culture and entertainment that normalized their presence in civilian life.[42] This played out as a sort of ubiquity for everyday civilians of both military hardware like drones and military practice in visual and digital culture across the news, on film and television, and of course in video games. Even as they normalize the presence of what Caren Kaplan and Derrick Gregory refer to as "everywhere war" and push these modes of engagement from battlefront to home front via policing and the securitizations of everyday life, such forms of "militainment" also expose the complexity and contradiction inherent in their logic.[43] The opening the sequence of the 2012 blockbuster *The Bourne Legacy*, for example, provides an apt illustration of how the types of paradoxical engagement that I am exploring played out for ordinary soldiers. The latest installment of the series and its most clear attempt to turn the original trilogy into a full-fledged franchise, the film introduces Alex Cross, the successor to Matt Damon's Jason Bourne character, via a series of super- or even suprahuman achievements as he traverses the Alaskan wilderness. As he fords glacial streams naked, scales craggy peaks, and fights off hungry wolves, the sequence sets up a classic man-versus-nature conflict only to demonstrate repeatedly the inherent superiority of man, or at least this particular man.

We eventually discover that this series of trials is a training mission designed to test the capabilities of the US government's newest weapons system: a line of chemically and genetically engineered soldiers. When a series of leaks threatens to reveal the program's existence, the CIA is forced to eliminate it by killing off the various members of the program. Given Cross's remote location, a drone strike is ordered to take out the final member of the arsenal. Demonstrating once again his cognitive and physical superiority, however, Cross manages to hack the tracking-chip technology embedded in his thigh by cutting it out and feeding it to the aforementioned wolves, one of whom becomes the unwitting decoy eventually executed by the drone. The scene offers a series of engaging role reversals in the hunter/hunted binary as Cross manages to turn both "predators" against one another by playing on their desires and limitations. With the introduction of the drone, a third term is inserted into the man/nature binary already established. As Cross fends off the instinctive, energetic desire of the animal with superior rational planning, he finds himself hunted on the opposite end of the spectrum by the calculated, quantitatively precise designs of the machine. Through a uniquely human combination of both tendencies, Cross is able to play animal and machine against one another.

This scene from an otherwise forgettable movie sets out an apparent conflict at the heart of the US military's application of virtual training and robotic weapons systems—one replicated in popular and scholarly discussions on the use of games and game technology in the military. On one hand, the military uses gaming and simulation technology extensively to train and recruit soldiers, hoping to channel the instinctive drives around popular first-person shooters into better-prepared soldiers. On the other, it deploys robotic weapons systems like drones that utilize telepresence to project force from the cool, rational space of the screen to the battlefield. But are these two forms of screen warfare actually connected? While a number of material connections and superficial resemblances present themselves, many soldiers and scholars have argued against simply equating them, a warning I echo in the discussion that follows.[44]

My goal in connecting them here lies in examining the range of nonfiction screen technologies deployed in the military's efforts during this period. While video games would seem to fall outside of the realm of nonfiction, I argue that their use in training and recruitment places them in a long history of military cinema. This placement aligns the cutting-edge efforts in *America's Army* with the more sedate and straightforward moving images the military produced as far back as 1898 and has continued to produce in vast quantities since the Second World War.[45] Drones, meanwhile, turn on the similar issues of embodied conflict and projected force, mediating the point of engagement to the degree that it creates a similar paradoxical relationship between the humans at the screen and the humans "on the ground." Though not identical, both present an iteration of the body/technology/space concern that *Gone Gitmo* manifests, an iteration that demonstrates bizarre hybrid forms that nonfiction media take on in the war on terror.

The military's use of games and other virtual technologies is a well-documented and discussed topic in game studies scholarship—from landmark texts by lay authors, like Ed Halter's *From Sun Tzu to Xbox*, to more-focused articles and chapters by leading scholars in the field such as Ian Bogost and Alex Galloway, not to mention anthologies of essays such as *Joystick Soldiers*.[46] As Halter convincingly argues, there have been game versions of war like chess for as long as there have been both games and war. The interconnections between the two take a significant turn in the 1980s when the military begins to express an interest in the nascent market of consumer video game technology and its possible use for training and recruiting. The starting point for what James Der Derian and Roger Stahl call the "military industrial media entertainment" complex or the "militainment" complex extends back over thirty years to the US Army's commission of Atari to modify its popular 1980 3-D vector game *Battlezone* into a training tool for future tank commanders.[47] The project produced *Bradley Trainer*, a stand-alone console game similar to the commercial version that was also produced, but modified to match the equipment profile of Soviet tanks. Though there is no evidence the project was ever used, it set an important precedent for the possibility of future collaboration between the commercial game industry and the military.[48] The military went on to sponsor adaptations of other popular games, including a customized modification of the game *Doom*, a popular online multiplayer game that allowed users to create custom environments. The military version, *Marine Doom*, could be unlocked by users with a special cheat code. In the late 1990s, the army entered into a partnership with the University of Southern California to form the Institute for Creative Technology (ICT). Intended to bring university-level research together with Hollywood creativity and military funding, the group produced *Full Spectrum Warrior*, a training tool that was also eventually released as a standalone console game.

The logic behind the military's use of commercial games technology appears synergistic on multiple levels. As Tim Lenoir and Luke Caldwell demonstrate, the combined efforts of the military and the game industry represented more of a fluid pool of collaboration and personnel as the institutions jointly conceptualized and argued for a radical reworking of the military's utilization of digital technology known as the "revolution in military affairs," or RMA.[49] The RMA was part dream, part blueprint, but with the impetus of 9/11 it began to move quickly off of the drawing board and into production. Collaboration before and after this transition offered a level of market-driven efficiency where none previously existed. The military could share the research and development cost of games with private companies seeking to outdo their competitors and directly monetize their investments. Second, collaboration enabled the military to benefit from the widespread popularity of games. On one hand, this had the effect of normalizing warfare on a larger cultural level, but it also lent the military a level of cultural cachet that it sorely needed in an era of all-volunteer soldiers. On a final level, the collaboration seems to have worked because industry and military interests often align. Both were interested in pursuing the most "realistic" simulations possible from both a visual and a mechanical perspective: the military, in order to prepare its soldiers for combat; and industry, in order to outdo its competitors in the technological arms race that defines the medium.

The military's use of game technology accords with Ian Bogost's influential argument about the procedural rhetoric of the game medium.[50] By invoking proscribed choices and actions on the part of players, games can procedurally persuade players that a certain idea, ideology, argument, or course of action is the preferred method for achieving a specific end. As Bogost demonstrates, the apotheosis of the military's use of game technology is perhaps the army's enormously successful and widely discussed title *America's Army*, which debuted as a free download in 2002 and has generated millions of downloads worldwide. Widely praised at the time for its high-resolution graphics and realistic simulation of battle, *America's Army* went through several other iterations, including the $12 million Army Experience Center and a stand-alone arcade game. Unlike the PC version, both of these versions allowed users to utilize gun-shaped controllers and other props to engage in combat.[51] The Army Experience Center in particular offered the general public a look at the more advanced technology the military had developed to prepare soldiers, including resources like the Infantry Immersion Trainer at Camp Pendleton, a thirty-two-hundred-square-foot facility that replicates "the sights, sounds and smells" of urban combat using a combination of physical settings and virtual avatars.[52] In this sense, *America's Army* is the tip of the spear, so to speak, in the military's strategy to enlist and equip soldiers for the realities of war. It's the first point of contact that many will have with what will eventually be a series of experiences using virtual immersive technology.

FIGURE 4.2. *America's Army* everywhere. The success of *America's Army* after its debut as a downloadable PC game persuaded the military to offer access to the game in public places and arcades. This enabled recruiters to approach players of the games directly, rather than waiting for them to enter their personal information on the website. It is worth noting that the arcade game (left) displayed signage that claimed the game was "Suitable for all ages." Photo credit: Carrie McLeroy (SMC—Army News Service).

It is worth noting that the military's extensive investment in and utilization of games and virtual environments to recruit, train, and indoctrinate soldiers into the ideology and procedures of military life was not its first interaction with media. For much of its history, the military used nonfiction film as its primary media technology for achieving many of these same ends. As Noah Tsika demonstrates, the military during and after World War II innovatively utilized documentary in a broad array of domains, pushing it formally and conceptually into service as a form of "useful cinema" capable of recruiting and training soldiers, documenting practices and tactics, and justifying military expansion.[53] Douglas Cunningham's work further elaborates the extent to which nonfiction training films within the United States Army Air Forces helped establish an esprit de corps within this new branch of the military and push the bounds of masculinity within the military more generally.[54] Even the treatment of psychiatric disorders and battle-related trauma or PTSD for returning soldiers fell under the purview of nonfiction film, becoming a tool to "recognize, diagnose, and treat the psychological effects of war," as Kaia Scott so brilliantly demonstrates in her history of World War II trauma.[55] As it would with *America's Army* during the war on terror, the military during previous wars utilized the medium of nonfiction film to round out its soldiers' "circuit" of service: enticing them to enlist, training them to fight, and dealing with the traumatic effects after they returned home.

Unlike previous efforts to meld games and military procedure, *America's Army* was engaging enough to attract a wide audience, while at the same time faithful enough to the nature of military conduct to constitute a valid training tool. All players have to adhere to the military code of conduct or risk being locked out of further play. Moreover, success depends upon cooperating with other players on group missions against enemy forces using the stock military equipment the game offers—all efforts to make the game as true to the army experience as possible. The game also includes a number of other nods toward the real army, including profiles of soldiers and stories of their time in combat that are featured on the game's loading screens.[56] The game faithfully reflects the larger ideological aims of American unilateralism and militarism, in that player/soldiers are constantly being deployed around the globe in an endless series of missions—a further level of verisimilitude between the real army and its virtual representation in the game.[57]

But even beyond these connections and references to the military, the game's training function further closely resembles the military's historical use of film to prepare its soldiers. A closer look at the format of the training sections of *America's Army* forcefully demonstrates the game's resonance with documentary practice. From its earliest iterations, *America's Army* always featured a notorious, or at the very least, onerous, training component. While most games are content to provide the basics of the controls or keyboard commands and assume that players will improve as they play, *America's Army* treats this instructional work as part of the end in itself. Throughout all of its versions, the game requires players to complete various exercises prior to being eligible to join in the more popular, team-based, multiplayer missions. In *America's Army 2: Special Forces*, for example, the training section includes five different levels. These stretch from "Basic Training" through various sections for weapons training, medic training, sniper training, and airborne training. In order to access the more advanced features of the game (including special weapons, access to certain maps, etc.), players are required to complete each of these exercises. At each stage, the game imposes a skill threshold that prevents players from progressing until they have achieved a specific score or level of competency.

For example, the "Medic Training" section presents several different subsections, including modules on controlling bleeding, airway management, treating shock, and so on. In each, the player appears in a hospital setting, complete with a reception counter and two medical staff people chatting about military life. The player proceeds to different classrooms that branch off of a central hallway, finding in each a classroom setting, an open seat, and a handout on the table. At the front of the room, an instructor stands near a projected PowerPoint presentation that leads the class through the given topic. Each lasts between five and ten minutes, concluding with a written, multiple-choice test contained in the handout. The slides that are used to deliver the content in these courses use photos, illustrations, titles, and bullet-pointed lists; in other words, they have all of the trappings of a typical slide show. In some of these, photographs of injured soldiers on the screen provide a pointed connection with (and contrast to) the forms of violence and physical injury that the game enables players to experience virtually. The appearance here of real bodies alongside the virtual or animated bodies of the avatars underscores the extent to which these sections of the game push closer to a level of reality than the other sections of the game.

While not alone in allowing players the ability to "practice" in order to advance their skills or in offering "tutorials" that introduce them to the mechanics of a specific game, the training sections of *America's Army* clearly exceed the practical requirements of game play. This is borne out in the falloff between the level of detail provided in the training sections and the practical execution and utilization of these skills in the mission sections of the game. Players who have passed medic training gain the ability to treat injured team members in order to bring them back into active play. But after learning in some detail how to treat basic wounds and manage a variety of common injuries, the game reduces these skills to a single command or click to put them to use. On the PC version of *America's Army: Special Forces*, players face the injured solider and press "E" for a specified period of time. In later iterations of the game, this is augmented with an on-screen representation of the avatar pulling out and applying what appears to be a roll of gauze. In essence, players are required to sit through nearly thirty minutes of training on basic first aid in order to gain the ability to press a single key on their keyboards.

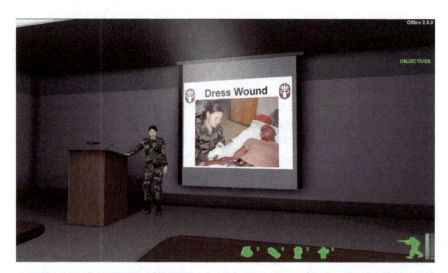

FIGURE 4.3. The "Medic Training" section in *America's Army,* version 2.5. The training segments, which last up 10 minutes each, replicate the experience of attending a lecture in a hospital, complete with an exam at the end.

Part of the explanation for the deeper level of engagement in this part of the game is of course ideological. The game's overarching goal is to portray the US military in an appealing light to the players who populate its servers. The end-game for a portion of these players is enlistment in the military, and for the rest it is presumably a positive opinion of the American armed forces. Greater detail here feeds into the larger world of the military that the game is introducing (or perhaps building), allowing players to immerse themselves in multiple dimensions of military life.[58] Thus, the idle chatter of the two staffers in the hospital hallway, the posters on the walls of the classroom, and ultimately the detailed instruction of the training tutorials themselves all further enhance what Galloway refers to as the game's "realistic-ness," if not its realism.[59]

Nonetheless, the level of information provided exceeds even this ideological, world-building demand. Instead, these sections seem intent on actually training players in, or at least introducing them to, the topics they present. As Lenoir and Caldwell note, the game's creators were insistent on its capacity to deliver a fun experience while at the same faithfully representing the military's structure, rules, and so forth.[60] This representation is gestured to throughout the game, but it appears most clearly at certain moments like those in the training sections. Here the game occasionally achieves something closer to nonfiction than its more action-driven components. While the embodied and affective experience of players during one of the game's combat missions surely departs radically from the experience of facing these situations in real life, the same cannot be said of the game's classroom-based training scenarios. I would argue that the experience of sitting in front of a computer and watching a lecture on a screen in preparation for a test is far closer to the experience one might have in real life doing these same things. This is certainly borne out in the complaints that the game generated for requiring players to endure these exercises. In this sense it would seem that the game is using the training sections to introduce players not only to the types of things they will be trained in should they choose to enlist, but also to the discipline of training itself.

The game's faithful simulation and enactment of a high degree of procedural and graphical realism for players does not, of course, obscure its more obvious distortions of reality. Some critics point out that the game's hyperrealistic images and soundtrack don't extend to the less palatable aspects of war, while others such as Galloway point out the extent to which the game exists in a sort of apolitical realm completely divorced from the social realities in which modern warfare is executed.[61] Indeed, at a most basic level, the game engages in the same "save-die-restart" logic that Nick Dyer-Witheford and Greig de Peuter call the "big lie of the video game as war" model.[62]

But there is an even bigger lie at work in the game, or at least a comparable one. To what extent does the game actually prepare its players for what is arguably the defining psychological reality of warfare: a kill-or-be-killed confrontation with another person? Of course, nothing entirely prepares one for that particular reality, which is the prime reason for training and simulation in the first place. Practicing "procedure" repeatedly prepares one to act on instinct when the time comes. This same logic also justifies the drive toward increasingly immersive environments and ever more realistic graphics that seems to characterize the game industry in general and its "militainment" branch in particular. It is certainly at work in the outsize investment the military makes in creating simulators for high-risk duties like flying a plane. The more time one can spend proximally adjacent to war without actually facing its mortal realities, the better. This is why the army apparently claimed at the outset of its participation in the ICT, "We want a Holodeck."[63]

And yet, as the military continues to modernize its arsenal with the type of digitized, networked technologies that brought *America's Army* to computers everywhere and further decreased the distance between simulation and reality, its simultaneous deployment of robotic technologies works to erase that distance entirely. The connections between games that simulate war and the robots actually used to fight it are both real and imaginary. On a cultural level, popular fictional texts that experimented with the possibility of playing a game while actually fighting a war began appearing shortly after the army and Atari teamed up on *Battlezone*. These included books like Orson Scott Card's *Ender's Game* and films like *The Last Starfighter, War Games,* and *Cloak & Dagger.*[64] In all of these texts, characters pursuing mastery of a game eventually learn that they are actually fighting a battle, thereby playing out the possibility that simulators like *Bradley Trainer* might eventually be linked to remote robotic tanks. Ronald Reagan himself prophesied that video games would eventually train a generation of soldiers to fight in war in "ways that many of us who are older [couldn't] fully comprehend."[65] As these technologies have become a reality, drones in particular and games like *America's Army* are routinely linked in press coverage about the emergence of "virtual war." And beyond these cultural expressions, Peter Singer notes that many of the controls for future robotic weapons are prototyped and eventually designed using hacked versions of the physical controllers and graphical user interfaces from console systems like the Xbox and PlayStation. This consistency is intended to limit the learning curve for the generation of soldiers who grew up playing games, but it also contributes to the slippage that exists in the minds of many between the games that recruit and train soldiers for war and the weaponized drones that are used to fight them.[66]

Outside of these connections, there is of course one important difference between the game's simulation and the drone: one is not real and the other is. Simulations prepare the player for a potential face-to-face violent encounter in their future; drones allow the operator to kill another human from a hitherto unimaginable distance. The expansion of this distance, as Gregoire Chamayou argues, is in fact the defining feature of the weaponized drone, one that turns it from fighting into hunting.[67] If games and training simulations are the cinematic equivalent of the blockbuster fiction film, then remote robotic warfare like the drone is its documentary equivalent. The parallels between the two are more than metaphoric. Like watching a big-budget action film, part of the pleasure involved in playing a typical action game is precisely the fact that it's not real, allowing one to engage in experiences and behaviors one would avoid in one's everyday life. This is what the "media effects" argument about the priming possibility of violent video games seems to miss.[68] Most people who avidly play first-person shooters do so because they *aren't* actually killing people. This is perhaps why so many soldiers, including drone operators, report playing these types of games in their free time to relax, as Peter Asaro notes.[69] Fictional games, like fiction films, allow us to experience realities that are thankfully not our own.

But like a documentary, drone warfare bears a necessary, mediated connection to reality. As Lisa Parks argues, the drone itself is a mediating technology that "exceeds the screen and involves the capacity to register the dynamism of occurrences within, on, or in relation to myriad materials, objects, sites, surfaces, or bodies on Earth."[70] I am even tempted to say an indexical connection to reality, if we bear in mind that one of C. S. Peirce's original examples of the indexical sign was the scar—evidence that signifies a prior wound due to some trauma imparted by the real. This index-as-scar is a connection more than borne out by the asymmetrical wounds experienced by both the operators and the targets of drone technology. Or perhaps more temporally and spatially accurate, his example of smoke to signify a distant fire.[71] To paraphrase Bill Nichols, no matter how realistic, games and fiction films will offer us only "a world"—as opposed to the world.[72] In the same way that viewers of a documentary are constantly aware that what took place before the camera bears a connection with the real, historical world, so the pilot of the aerial drone knows that when he pulls the trigger, an action and effect are carried out in the world.

Beyond the presence or absence of a direct connection to reality, there are other parallels between these forms of media. Like state-of-the-art special effects in fiction films, games and simulations utilize cutting-edge technology to achieve photorealistic visual imagery. By comparison, the typical camera image available to the drone pilot can have the feel of a low-res, low-budget documentary shot on consumer-grade home video. Even as camera technology and bandwidth have expanded to include infrared, higher-resolution imagery and multiple simultaneous angles (the "gorgon stare" model can track twelve independent locations and quilt these together in a single, unified field of view), these tools are aimed at revealing rather than simulating reality.[73] As opposed to the constant action of a game or simulation that allows one to skip to the "good parts" of the text, one of the prime problems confronting drone pilots is the boredom induced by extended hours of inactive screen time—a fear not uncommon to film majors taking their mandatory documentary course.[74]

Considering drone warfare in the context of documentary further alerts us to the heavily mediated nature of the reality experienced by the drone pilot, as well as the clear benefits and drawbacks of experiencing reality from a "safe" distance. As has been well publicized, drone pilots often act on less-than-sufficient information, resulting in tragic civilian deaths.[75] Derek Gregory's description of the "kill chain" demonstrates that pilots and sensor operators are just two of the dozens of people involved in a typical mission. This group communicates across a variety of locations and through multiple channels including voice, IRC (internet relay chat), and in person—a sociotechnical assemblage of humans, information, and communication channels that presents multiple failure points.[76] Many estimates put the number of civilian deaths as compared to combatant deaths well into the double digits, meaning dozens of innocent people die for each individual targeted by this class of "precision" weapons.[77]

Even the pilot and sensor operator, though admittedly removed from the direct encounter of the battlefield, are hardly out of harm's way entirely. Making life-and-death decisions from an air base thousands of miles away in northern Nevada or Virginia and then driving home to the suburbs when a shift is over has been blamed for causing PTSD in drone pilots at a rate comparable to that of front-line soldiers. Even soldiers who don't experience the acute symptoms of PTSD suffer from a high rate of burnout, owing to an overwhelming set of labor conditions.[78]

While higher-resolution optics and more-precise strike capabilities are being planned to alleviate the problem of mistakes, such measures would seem only to exacerbate the circumstances that give rise to the psychological conflict associated with PTSD. Pilots and sensor operators report that watching the same person for an extended period of time only increases the guilt and anxiety that arise when they eventually kill him or her.[79] As in documentary film, no particular optic or recording technology can reconcile one to the larger moral and ethical "truth" of what one is seeing or doing.

Most importantly, however, comparing the military use of virtual and robotic technologies to fiction and nonfiction films throws into relief the connected but different nature of these mediated forms of fighting. Like fiction and documentary, games and robotic warfare exist on the same spectrum of representation but cannot be conflated. Moreover, comparing the experience of drone pilots with the promise offered by something like *America's Army* clarifies the bait-and-switch effect at work in the military's deployment of virtual technology to recruit and train soldiers to fight its wars. Whereas the game offers players an enhanced sense of agency, excitement, and immortality, war as experienced by drone pilots seems to entail guilt and boredom—experiences that are absolutely anathema to the feelings that commercial gaming, and by extension *America's Army*, are supposed to evoke. It is war fought from the safety of one's home, but it turns that home into a battlefield where the "combatants" have to balance soccer practice and family dinner with killing people. Indeed, such distinctions (between civilian and soldier, battlefield and home front), always tenuous at best, are only increasingly difficult to separate in the particular conflation of space that drone warfare invites. To return to the example of Alex Cross in *The Bourne Legacy*, the military wants to place modern soldiers somewhere between the instinctive, hedonistic experience it uses to entice and train them and the sanitized, strategic space of surgical, robotic combat. And just as Cross's superiors do in the film, it has forgotten to take account of the human in the middle.

But if the military has opened up a curious new "front" in combat through drones and the bending of space that they enable and has further sought to colonize virtual technologies and the spaces they create in order to recruit and train the bodies it sends to fight the war on terror, it has also opened up a space of resistance that artists and activists have worked to engage. While my focus here has been on considering the military's use of both drones and games vis-à-vis the soldiers it engages through each, I would like to briefly consider a few of the particular interventions that have sought to counter the military's particular applications of both technologies. The artist Joe DeLappe in particular has worked across many of these same spaces with projects that, as the curator Christiane Paul puts it, "expanded, challenged, or even redefined concepts of what constitutes public space, the public domain and public art."[80] In Second Life, DeLappe adapted a treadmill to work as an interface control and modified his avatar to resemble Mahatma Gandhi, walking across the virtual landscape as a reenactment of Gandhi's "Salt March."[81] In 2006, DeLappe decided to engage the *America's Army* platform by utilizing the game space as a site through which to memorialize the real soldiers who had died fighting in the war in Iraq. The *Dead in Iraq* project consisted of logging in via his screen name ("dead-in-iraq") and then dropping his weapon once the game commenced and using the game's chat function to manually type the name, age, service branch, and date of death of each service person who had died to date. Eventually, often quickly, his character would be killed, and the other players would respond to the intervention. Over the course of five years, DeLappe logged in hundreds of times, finally completing the list of all 4,484 names in 2011 after the US Army had officially withdrawn from Iraq. Many of the chat transcripts from these sessions reveal a mixture of reactions, ranging from sympathy and sadness for the soldiers who passed away to anger and annoyance that he was confusing a game with reality, or spoiling the fun.[82] Conceived of as a "fleeting memorial" for these deaths, their names populate the fictional space that was created in order to recruit and train their successors.

In 2014 DeLappe began a series of projects intended to challenge the space that drones occupy within spatial imaginaries of the countries that most often deploy them. DeLappe's drone work ran across a heterogeneous array of spaces and interventions, stretching from the "personal drones" that he created for individuals to wear as headbands to the "In Drones We Trust" stamp that he created for users to stamp the back of US currency with a small image of an MQ-1 Predator, thereby placing the drone's silhouette over some of the most celebrated American landmarks. Other interventions include placing scale-model drones in public spaces and modifying existing paintings with a drone image similar to the one that now populates an unknown number of bills. Connecting the projects is an impulse to counter the drone's tendency to populate the periphery of our minds, drawing it from the extraterritorial "borderlands" where it terrorizes people in our names and allowing its shadow to be cast on the minds of a population all too willing and all too able to forget about it.

CONCLUSION

Side by side, little separates the form and methodology of *America's Army* from *Gone Gitmo*. Aesthetically, they operate within the same register: a virtual landscape or object on one platform looks as "real" as a similar object or landscape on the other. While both trail behind the most cutting-edge virtual representations available, they both nonetheless achieve sufficient verisimilitude that we recognize objects and places for those they are supposed to represent. A fence is quite obviously a fence. Both also utilize these virtual environments to offer users a sense of what it would be like to inhabit a real space quite different from their own. Where *Gone Gitmo* seeks to offer visitors to the virtual Guantánamo a sense of the physical environment and the political and legal issues facing the detainees, *America's Army* wants to prepare future recruits for possible events they might face as soldiers engaged in combat. Both also draw upon traditional photographs and video footage as ancillary materials to point users back to this reality in a similar fashion—*Gone Gitmo* through the video testimony that it includes from detainees and *America's Army* through its "Real Heroes in Action" profiles, segments on the game site that feature profiles of decorated soldiers.[83] Following from these formal similarities, both have recourse to one of the defining features of persuasive rhetorical media, documentary film included: the presence of a clearly defined call to action that structures the text. *Gone Gitmo* aims to stimulate debate and protest around the ongoing imprisonment of the individuals detained at Guantánamo. *America's Army*, on the other hand, primarily hopes to persuade young people to enlist in the US armed services and, more indirectly, to shift general opinions of the military more broadly.

And yet in spite of these similarities, the above reading demonstrates that one sits closer to activist documentary while the other lies closer to fiction film. The spatial metaphor is important here, as distinctions between fiction and nonfiction, real and virtual, are better understood by degrees of separation and resemblance than by categorical distinctions. The comparison between these forms of media illuminates the importance of holding on to the (admittedly slippery) categories of fiction and nonfiction even as the technological ground shifts away from optical recording technology. Just as optical indexicality never guaranteed documentary film's relationship to truth, nothing in the graphical resolution or three-dimensional rendering technology itself guarantees any closer or more complete relationship with the events that each seeks to represent.

Nor does the distinction between the two texts rest on their differing political orientations (one apparently opposing the official status quo of military policy, the other supporting it). While *America's Army* has been dismissed by its detractors as propaganda for the military, the designation mischaracterizes the relationship between a politically persuasive text and a propagandistic one. This is of course a vexed issue, and any working hierarchy between the two reveals more about a given value system than it does about either form's relationship to reality. We tend to believe in truth claims that support our individual ideological frame and dismiss those that contradict it. "They" make propaganda; "we" speak the truth. Reality always suffers any number of distortions when forced into particular narrative frames. Gaps and omissions are the rule for both what we dismiss as propaganda and what we deem "true" to reality.

The tipping point between fiction and nonfiction in traditional and virtual documentary lies in what these gaps and omissions exclude or include, and the extent to which we deem these choices to be critical to the project's larger truth claims. *Gone Gitmo* excludes abstract or embodied experiences like indefinite detention and torture, both of which might be considered "unrepresentable" in a broad sense. Indeed, both are practical impossibilities given the nature of the chosen medium; thus the project invites participants to reflect on such realities by way of their obvious omission. In *America's Army*, the gap in the text is instead the defining feature of the chosen medium. That is, the first-person shooter is arguably defined as a forum in which the player is able to engage in consequence-less killing firsthand, with no corporal, legal, or ethical jeopardy at stake. Warfare, on the other hand, is the exact opposite. *Gone Gitmo* chooses *not* to represent these things, but to instead play on their absence. *America's Army* utilizes this representational distance to make killing ubiquitous and individual death a mere inconvenience. (One has to start the game over.) This distinction makes one a documentary and the other a fiction.

The fictional status of *America's Army* is further underscored by the mediated experiences of drone pilots—soldiers who actually fight in combat situations but do so through a gamelike interface. Unlike virtual representations, the video and data feeds that confront pilots don't point toward reality; they emanate directly from it. This is still reality represented, but done so without an author, seeking not a broader "call to action" but instead a stimulus/response from the soldiers who jointly monitor its various feeds and collectively weave together the bizarre text of a drone mission. The attendant feelings of stress and guilt that such experience seems to evoke for many of these individuals further highlights the lack of stakes in the game version of virtual war, which allows and even rewards higher body counts. While the experience of remote, telepresent combat that the drone interface provides offers important parallels to documentary representation, there are clear distinctions that make a direct connection between them problematic.

In its efforts to curb the incidence of PTSD for drone pilots, the military began investigating technology to give the computer systems that pilots and sensor operators interact with more personality.[84] The thought is that sharing the guilt with a "third party" would lessen the burden shouldered by the individuals pulling the trigger. Depending upon the level of automation the military eventually achieves, a fully automated drone fleet might alleviate the need for human operators entirely. While this would further obscure the visibility of the program from the public in whose name this warfare is waged, it would also necessitate the use of new technologies by activists like Weil and de la Peña seeking to make its absent reality more tangible to everyday citizens, even if only through virtual means.

5

Technology, Transparency, and the Digital Presidency

Sunlight is said to be the best disinfectant.

—LOUIS J. BRANDEIS, *OTHER PEOPLE'S MONEY*

Science and art have in common intense seeing, the wide-eyed observing that generates empirical information. [It] is about how seeing turns into showing, how empirical observations turn into explanations and evidence.

—EDWARD TUFTE, *BEAUTIFUL EVIDENCE*

On March 28, 2011, a group of representatives from five different government watchdog groups met at the White House to present President Barack Obama with an award in recognition of his efforts toward creating greater government transparency.[1] The presentation was intended to coincide with an annual event known as Sunshine Week, which has, since 2002, sought to raise awareness about greater access to and oversight of the government by the press and individual citizens.[2] In an odd public-relations gaffe, however, the White House chose not to make the presentation an official press event and consequently held the meeting in private. As perhaps should have been expected, the press immediately jumped on the irony of the situation, and for the next twenty-four hours, headlines like "Obama Accepts Transparency Award . . . in Private!" appeared across the media. While the award's presenters were critical of the discrepancy between the event and its public profile, they nonetheless reiterated their praise of the president's efforts to make the federal government more open, and pushed him to continue in his pledge to make his presidency the most transparent in history.

In spite of the absurdity of recognizing openness in private, however, a deeper irony underscoring the award seems to have gone unnoticed by all involved. In the spring of 2011, the US government was perhaps more open than at any point in its history, but it wasn't just Obama who deserved the credit. Although his administration strove to overhaul policies and procedures in order to push mountains of data into the public domain, its efforts were overshadowed by a series of high-profile unofficial leaks by the organization WikiLeaks. Combined, the 2011 WikiLeaks releases placed more than a million previously secret documents online. Alongside the surveillance programs uncovered by Edward Snowden in 2013, the leaks opened up a heated debate about the need for secrecy as well as transparency in public life. While Obama's policies were meant to signal a shift from the secrecy that had characterized his predecessor's administration, the WikiLeaks scandal forced the administration to admit that there were limits to the degree of openness it was comfortable with.

This chapter explores the threads of openness and transparency as they are woven into debates around the high-profile "data dumps" (both official and un-official) that emerged from US government archives during the first years of the Obama administration. Although government transparency has long been con-sidered an ideal within democratic politics, different forms of media—from news-papers and photographs to film and television—have been celebrated as the best means of achieving this goal at various points. It was clear from the start that for the Obama administration, transparency and digital technology went hand in hand. Out of this marriage, data visualization emerged as the favored medium to carry out the administration's ambitious proposals, creating a new form of digi-tally driven documentary media in the process. WikiLeaks shared Obama's belief in transparency and his faith in technology as the means to achieve it, but from an anarchic, oppositional position. Given the size and scope of its releases, journalists and amateurs alike also turned to new graphical tools like data visualization in order to mine and display the WikiLeaks data.

Long used to represent scientific and financial data, data visualization, with its ability to represent vast quantities of information at a glance, offered an ideal medium to capture the government's complex inner workings in a legible way for the general public. While it may seem that data visualization subverts the repre-sentational and indexical media that previous chapters considered, image-based media continue to reassert their importance and influence. Not only do still and moving images continue to appear alongside data visualization's interactive charts and graphs on government websites, but, as its evolution will demonstrate, data visualization's claim to representing reality rests on the same "seeing is believing" foundation that photographic media first established. Data visualization is an out-growth and inheritor of the same documentary impulse that has for many decades fueled the production of nonfiction film and television in general.

While the US government and WikiLeaks were at odds over the nature and purpose of releasing information to the public, both were nonetheless part of a larger historical debate over the definitions of "open" and "transparent"—one geared toward expanding notions of publicity and public space. Echoing Louis Brandeis and other Progressive Era reformers from a century earlier, these new champions of transparency and government accountability believed that "sunlight is the best disinfectant" for clearing away the miasma of government corruption and public distrust. Unlike their forerunners, this new generation of transparency advocates believed that the best technology of visibility was not the camera or the muckraking exposé, but instead the networked computer—a tool capable of visualizing and distributing the vast amounts of data being made public through official and unofficial channels. Combined with a Depression-era belief in the power of documentary evidence to inform the public, groups on both sides of the law worked to usher in an era of transparency and accountability akin to early- and mid-twentieth-century efforts, but this time with a distinctly twenty-first-century twist.[3]

DATA VISUALIZATION: THE ROOTS OF
A REVOLUTION

The use of data visualization by both the Obama administration and WikiLeaks to make their transparency efforts legible to the public should perhaps come as no surprise. Data visualization as a medium is currently in the midst of an immense level of popularity and prominence across contemporary visual culture. Examples can be found everywhere, from advertising (IBM's "Smarter Planet" campaign)[4] to journalism (CNN's "Magic Wall" and NYTimes.com's interactive infographics)[5] to academia (the journal *Nature* and others, as well as the burgeoning field of analytics). Visualization is celebrated on blogs like informationisbeautiful.net and visualisingdata.com and made available to the masses via tools like Wordle, Tableau, and Many Eyes. Even as they struggle for added revenue amid falling circulation rates, newspapers like the *New York Times* and *The Guardian* have invested heavily in the medium. They regularly offer online features and interactive visualizations not possible in print editions, fostering the new field of "data journalism" in the process.[6] Visualizations are so ubiquitous that as early as 1982, in a pre-Internet era of publishing, Edward Tufte was able to claim, "Each year, the world over, somewhere between 900 billion (9 X 10^{11}) and 2 trillion (2 X 10^{12}) images of statistical graphics are printed."[7] This staggering figure might still be rivaled by the number of photographic images captured every year by the increasingly ubiquitous camera-enabled smartphone, but it's the contention of this chapter that the two should be thought of as being in the same category of media.

As a contemporary phenomenon, the recent explosion of data visualization relies on a range of newly emergent technologies, from the widespread deployment of inexpensive sensors to collect data, to advancements in cloud computing and commodity-level clusters for processing and analyzing this data, to high-resolution displays and increased graphics processing for rendering it visible.[8] But in spite of its recent popularity in contemporary media and its reliance on cutting-edge technology, data visualization broadly conceived (the impulse to collect information about the world and display it visually) is among the oldest of pursuits.

EARLY HISTORY AND "GOLDEN AGE"

The current renaissance of computationally produced data visualization is only the latest iteration of an aesthetic form that stretches as far back as the first scientific tools and earliest forms of human writing.[9] Many contemporary data scientists and visualization researchers chart the roots of their work as far back as 6200 BC, to early maps found in Konya, Turkey, that demonstrate the desire to graphically depict the physical world.[10] This same impulse continues through Ptolemy's *Geographia* in AD 150 and reappears in various scientific and technical representations throughout the Renaissance, extending into William Playfair's experimentation with line charts and bar graphs in the early 1800s. Playfair's work stands at the beginning of what historian Michael Friendly refers to as the "golden age" of data visualization—a period during the second half of the nineteenth century when simultaneous developments in a range of fields fostered an explosive period of innovation and a volume of output rivaled only in the last fifteen years.[11] For Friendly, this period "deserves to be recognized—even revered—for the contributions that it made to statistical thought and practice in that time and for the legacy that it provides today."[12] As with our contemporary moment, these innovations touched on each phase of the transformation from data to graphics, including data collection (improvements in scientific measurement, instrumentation, and cartography), analysis (developments in the fields of statistics and demographics), and display (the invention of processes like lithography to print and distribute full-color graphics on a mass scale).

This golden age of infographics stems from a larger nineteenth-century revolution in science and mathematics, one in which new ways of observing the world were developed alongside new forms of processing and understanding those observations. The nineteenth century, as Ian Hacking argues, was bookended by the determinist models of Newtonian mechanics on one end and the open-ended indeterminacy of quantum mechanics on the other. Through this metamorphosis, the natural world came to seem "regular and yet not subject to the universal laws of nature."[13] Statistics, through the development of probability theory and what Hacking memorably describes as the "taming of chance," gained the ability to describe the world in regular, repeatable patterns. As in our current moment, data began to penetrate and influence the inner workings of social and political life in a way that it hadn't previously, touching on everything from medicine to policing to agriculture. Through the work of figures like Adolphe Quetelet and Charles Babbage, Joseph Fourier and Frédéric Villot, Charles Dupin and others, detailed records for virtually every aspect of the modern nation-state were recorded, tabulated, and of course graphically represented for the first time.[14] The rise of probability theory in statistics as a scientific model went hand in hand with the sorts of vast data collection and visualization that Friendly celebrates, resulting in what Hacking refers to as an "avalanche" of printed numbers. This avalanche provided the foundation for and entrenchment of the modern bureaucratic state, interpolating in the process a new view of the public that "has affected not only the ways in which we conceive of a society, but the ways in which we describe our neighbor."[15]

For Michel Foucault—pointing out as he did that "statistics is knowledge of the state, of the forces and resources that characterize a state"—the nineteenth century bound modern statecraft to statistics even more directly. Part of governing a population, then, relied upon the collection and administration of "its quantity, mortality, natality; reckoning of the different categories of individuals in a state and of their wealth; assessment of the potential wealth available to the state, mines and forests, etc.; assessment of the wealth in circulation, of the balance of trade, and measure of the effects of taxes and duties, all of this data, and more besides, now constitute the essential content of the knowledge of the sovereign."[16] Indeed, the rise of what Foucault has called "the sciences of man" (and the statistical models and visualization tools they utilize) plays an essential role in the state's administration of its power. Numbers, and their visual forms, enable states to control and regulate the flow of people and resources within and through their borders. As we will see when we turn to more-contemporary examples of data visualization and their role in the transparency debates surrounding the Obama administration, the flow of the "knowledge of the state"—who is allowed access, who is subject to it—remains at stake.

Statistical data collection and graphical representation thus moved from describing natural phenomena in the physical world (e.g., Galileo's records of the movements of heavenly bodies) to describing, and eventually influencing, the nature of the social and political worlds (e.g., Minard's celebrated map of Napoleon's ill-fated march into Russia). This shift signaled not only an aesthetic change in how data was processed and displayed, but also a conceptual shift about what types of phenomena could produce data and to what uses the information could be put. This move from representing the physical to the social world is perhaps the most important legacy our current moment inherits from its nineteenth-century roots. It is also one that places techniques of data visualization in league with two other modes of representation—namely, photography and film.

PHOTOGRAPHY AND DATA VISUALIZATION: TWIN HISTORIES

The formal and procedural similarities between contemporary data visualization and photographic media being outlined here trace their roots back to the historical coincidence of the emergence of both forms. About the same time that

William Playfair and others were pioneering the techniques and models that would initiate Friendly's "golden age" of data visualization, Nicéphore Niépce and Louis Daguerre were beginning the collaboration that would produce the first reliable method for chemically recording and reproducing images captured in a camera obscura–style device.[17] By the time François Arago arose in the French Chamber of Deputies to formally announce the procedure perfected by Niépce and Daguerre in 1839, its benefit to scientific practice as a "valuable aid" of calculation and observation in fields as far-ranging as astronomy, microscopy, and anthropology were already clear.[18]

In claiming photography as a scientific tool, Arago was arguing one side in the debate that seems to have dogged photography from its earliest uses: whether it constituted a genuine art form or merely a mode of technological reproduction. Notably, it seems to have gone unquestioned that photography was always, at least, a tool for science and scientific observation. The automatic reproduction of reality without the mediating hand of the artist or scientist rendered photography both suspect as an art form and ultimately useful to scientists. Photography became both a symbol of the standard of scientific objectivity as a whole and one of the tools by which individual scientific results were documented. Through its use in the laboratory, a certain "facticity of the photograph" was secured through "a distinct form of scientific comportment that harnessed photography to rigid protocols," as Robin Kelsey and Blake Stimson put it.[19] In tones reminiscent of contemporary accounts of the importance of data visualization for generating insight, scientists utilizing early forms of photography to stop motion and freeze time hailed it as capable of offering humans a power of observation their eyes did not have. Its incorporation into the laboratory setting spurred a shift in scientific observation and documentation, bringing with it a standard of what Lorraine Daston and Peter Galison refer to as "mechanical objectivity".[20] As data visualization would a century later, photography seemed to cast light on an aspect of the world otherwise hidden from human observation.[21]

In the late nineteenth century, the world was becoming representable and represented, not simply calculable but increasingly quantified and measured, in means beyond the written word and the painted image. Photographic media (both moving and still) and data visualization were the means through which this transformation occurred. As a medium emerging in and among the same social and scientific changes that gave rise to the "taming of chance" and the establishment of the world as statistically calculable, photography was energetically identified as one of the forces of modernity shaping social life. Oliver Wendell Holmes's much cited celebration of stereoscopic photography placed the automatic reproduction of images alongside the railway and the telegraph as an invention whose "significance forces itself upon us daily."[22] For Holmes, the exchangeability of the image for the referent and its freedom to circulate made it capable of "annihilating time and space—a potential many would later grant to film as well. Holmes's prediction of a time "when a man who wishes to see any object, natural or artificial, will go to the . . . stereographic library and call for its form" evinces the same desire for ubiquitous documentation of the world that drove the "avalanche of numbers" that Hacking and Friendly credit with the birth of data visualization during the same period.[23]

Given the close connection between scientific observation of the world and photographic documentation, it is unsurprising that the first steps on the path from still photography to motion pictures were undertaken by scientists seeking to perfect the still image as a means of observation. Étienne-Jules Marey and Edward Muybridge, the standard figures cited in histories of cinema, both stumbled on motion pictures while doing "other" scientific work. For Marey at least, it seems that motion reproduction, in the cinematic sense, was the least relevant by-product of his photographic work.[24] While he would later lament that the moving image's ability to capture "simply what the eye can see" rendered it useless as a tool of scientific observation, film's ability to accurately and automatically document the world cemented its place in the popular imagination as a substitute for physical presence in a given place.[25]

Almost from the moment they appeared, nonfiction moving images were deemed capable of documenting, preserving, and revealing the world. While documentary film in the standard definition of the genre wouldn't appear until the mid-1920s, from the Lumière brothers on, the impulse to document the world and consume the resulting footage has persisted in a variety of forms over the last century. Many of the same tropes from these early actualities reappear in the latest cell-phone footage on YouTube, including individual records like home movies as well as social/historical records like network news coverage of important events. As a tool, nonfiction moving images have always been utilized in the way scientists and data journalists utilize visualization today: to document and reveal the world to others.

Of course, like early forms of data visualization and the political and social frameworks they supported, photography and film have both been criticized for mediating and shaping the historical world as much as they record and reproduce it. Much of the critical work undertaken in the last fifty years has gone toward pointing out the lie behind photographic indexicality's seemingly untroubled connection with the physical world. A photograph is now thought to reveal as much about the photographer as it does about the subject—a by-product Kelsey and Stimson call its "double indexicality."[26] Moreover, as the work of John Tagg makes clear, photography itself was as much a part of the rise of the bureaucratic state and the normalizing of everyday life as was the data collection and statistical representation of the late nineteenth century. From the use of photography as a means of surveillance by police to the documentation of slums by governments seeking to rid society of their social ills, photography has played part in the power of the state equal to its more numerically driven counterparts: statistics and data visualization.[27] If we add to this tools such as closed-circuit television, satellite imagery, and other optical forms of state surveillance—tools that have created a panopticon of power within modern social life that affects everything from urban planning to individual behavior—then moving images require as much scrutiny as tools of state power as their still forerunners.[28]

PHOTOGRAPHY AND DATA VISUALIZATION: DISTINCT MEDIUMS

In spite of their shared histories and the similar roles they have played in scientific, social, and political contexts, photographic media and data visualization are nonetheless distinct media that offer starkly different modes of representation. The differences between the two become apparent by considering the role that each has played in relation to the transparency movement. Although photographic media have been used as tools of the state in its administration of power, it would obviously be misleading to conclude that photographic images have benefited or been utilized by just the government. As an optical medium, photography was a natural fit for those who wished to keep an eye on the state and reveal its misdeeds to others. Data and data visualization, on the other hand, have until quite recently remained the province of the state. One characteristic of the modern transparency movement is its clear desire to make data visualization as accessible as photography has been for the past century.

A long tradition of social-issue photography exists in the United States, stretching as far back as Jacob Riis's documentation of New York slums in the 1890s and continuing through the work of Lewis Hine, Dorothea Lange, Margaret Bourke-White, and Walker Evans in the early twentieth century and extending on through the work of photojournalists like Kevin Carter, James Mollison, and others today.[29] While charges of exploitation and patronizing paternalism have always been levied against such work, the connections between the photographic representation of social issues and the healthy functioning of representative democracy continue to inform the impulse to visually document and distribute certain images as a tool for social justice.[30]

The work of Lewis Hine, for example, is often considered to have played a major role in the implementation of child-labor laws in the early twentieth century and in the rise of labor rights more generally throughout the 1920s and 1930s. In words reminiscent of Tufte's ambitious claims for data visualization, Hine often claimed a utopian strength for the power of documentary photography. He wrote:

> The picture is the language of all nationalities and all ages. The increase, during recent years, of illustrations in newspapers, books, exhibits and the like gives ample evidence of this.

> The photograph has an added realism of its own; it has an inherent attraction not found in other forms of illustration. For this reason the average person believes implicitly that the photograph cannot falsify. Of course, you and I know that this unbounded faith in the integrity of the photograph is often rudely shaken, for, while photographs may not lie, liars may photograph. It becomes necessary, then, in our revelation of the truth, to see to it that the camera we depend upon contracts no bad habits. . . . The dictum, then, of the social worker is "Let there be light"; and in this campaign for light we have our advance agent the light writer—the photograph.[31]

Hine calls here on the full weight of long-standing Enlightenment connections between truth and social justice, drawing metaphorical and physical connections to the process of photographic exposure. In doing so, he tacitly channels the deep-seated liberal faith in the connection between the presence of information about a problem and resulting action toward social justice. This faith—that exposing a problem will result in its elimination—is one that runs throughout work on government transparency and political accountability, regardless of which medium is acting as the channel of transmission.

As Jonathan Kahana points out, much of the history of political documentary film in the United States rests on the notion that moving images can be used to achieve social change, whether this change is political accountability for those in power or greater justice for those without it.[32] Kahana covers a range of formal practices from the Depression era through the social crises of the 1960s and up to contemporary political work. He writes, "In an emancipatory gesture that [we] find repeated over and over . . . each style of documentary claims in its way to liberate its viewers from ignorance, prejudice, false consciousness, or illusion."[33] Indeed, much of the moving-image work considered in each of the previous chapters echoes this same gesture. From the overt skepticism of Errol Morris through the straightforward polemics of Robert Greenwald and on through the virtual walls of *Gone Gitmo's* Guantánamo Bay, each of the works considered uses a variety of formal approaches, but all share the basic aim of educating viewers on a set of issues. As a medium built on the same physical and photochemical principles utilized in still photography, political and social-issue documentary have inherited this faith in the connection between the light that exposes the film and the light that can expose social injustices—a spirit Hine so clearly articulated.

Turning to data visualization and the history of government transparency, we find a somewhat different history—one, in fact, that is far shorter. While notions of openness and transparency have been an ideal of democratic states from their earliest instantiations, the usual organ for achieving these goals has been the existence of the principle of freedom of the press. Access to government meetings and records at all levels by the press has been a consistent source of conflict, one in which degrees of access and publicity have varied throughout different countries and time periods. During the Progressive Era, for example, reformers like Louis

Brandeis and Woodrow Wilson called for greater information sharing on the part of the government in order to counteract the perceived corruption of politics by business interests. Brandeis's notion of "publicity" drew on his belief in the ability of the public gaze to root out corruption if given proper access to the information. Thus, one of the clear goals of Brandeis and others was making information known to the government available to the public as well.[34] Indeed, his well-worn observation that "sunlight is said to be the best of disinfectants" evokes metaphors similar to Hine's and directly connects access to information with optical visibility as a means to expose truth.

These early parallels notwithstanding, something like a comprehensive or even limited collection of information or records from and about the government that could be subject to analysis by independent groups and visualized for the public has emerged only within the last few years. This is in part the result of two factors that we have already seen in the history and development of data visualization. The first is that for much of its history as it relates to social and political life, data visualization has been a tool for the state to more effectively administer power. To be carried out, large-scale data collection and record keeping often required the resources of the state, and such records, moreover, were often kept secret by governments.[35] For example, one of the first uses of the early data-sorting machine that Herman Hollerith designed was in tabulating the 1890 and 1900 censuses for the US government, one of the only potential customers with a need and a budget to support the new technology.[36] Even as Hollerith's machine became the centerpiece of IBM and was eventually surpassed by digital mainframes and cloud computing, the company continued to rely heavily on sovereign governments around the world as the main customers for its cutting-edge computing technology. States were the only entities that had both the resources to purchase these technologies and sufficient quantities of data to justify them. But by the next centennial census, in 2000, technology like the Google MapReduce framework had made large-scale computing a commodity open to anyone capable of networking two or more computers together, and the ubiquitous presence of computers for the previous fifty years had created mountains of data that were increasingly available to anyone with an Internet connection.

The second change that enabled the current explosion of transparency-related activity was a further redefinition of what counts as data, and thus what can be visualized. In the same way that the golden age of data visualization was fueled in part by expanding the notion of data to include certain measures of social and political life, the current resurgence of the field is similarly expanding the idea of what falls within its purview. Part of the recent revolution in the field of data visualization, for example, has been the development of a related field known as natural-language data processing (NLDP for short). NLDP analyzes written texts and categorizes them according to various criteria, thereby generating data about them and hence the ability to visualize this material. Unlike conducting a census, where data must be collected and classified, NLDP allows one to turn any written material into a data source (not just books, but e-mail, websites, tweets, etc.). In political circles, the same army of bureaucrats it takes to conduct a national census also produces an overwhelming archive of written material in the process of everyday governance. Even though reforms in the United States like the 1966 Freedom of Information Act (FOIA) and the Ethics in Government Act of 1978 (passed in response to the Watergate scandal) have long given citizens access to this information, until recently such records had been made available only in printed form, thus requiring an equally sized army of individuals to sort and make sense of such information. The advent of digital record keeping and NLDP makes this archive a viable source of data that can be analyzed and visualized by anyone with relatively little computing power.

Though photographic media and access to government data have both been invoked utilizing Enlightenment metaphors of light, optics, and vision as a direct access to truth and justice, it is only in the last decade that information visualization became a widely available form of visual media. Photography has enjoyed this status since the early twentieth century thanks to consumer-grade equipment like the Kodak Brownie. Until recently, free access to information as an avenue toward government transparency typically meant access to printed records in a specific government agency during prescribed hours. Any public insight into what such records contained was usually provided via written reports from the press or a specific watchdog group. Once these records become digital, however, they become open to investigation by a much wider segment of the population using a broader array of tools for summarizing and accessing their contents.

THE REBIRTH OF DATA VISUALIZATION

The current body of practitioners, theorists, and researchers working on data visualization maintain a curious relationship with the medium's past. On one hand, the field demonstrates a clear awareness of its own history. Textbooks on visualization often begin with an introductory snapshot of successful examples and important milestones, each perhaps covering different periods but all invariably highlighting a fairly consistent canon of work. And yet, despite this historical awareness, the claim is often made that what unites all of these materials is a kind of timeless, universal aesthetic that appeals to an innate, almost biological aptitude for this particular mode of representation. As Edward Tufte puts it, "The principles of analytical design are universal—like mathematics, the laws of Nature, the deep structure of language—and are not tied to any particular language, culture, style, century, gender, or technology of information display."[37] Data visualization has a history, but it is simultaneously and consequently thought of as timeless.

At the heart of what connects early diagrams in Euclidean geometry with computationally rendered scientific charts and graphs in the latest issue of Nature is a long-held faith, in Western culture, in the apparent connection between vision and human cognition. As Martin Jay, W. J. T. Mitchell, and others working on visual culture have argued, the connections between seeing as one of the human senses and cognitive concepts like understanding, knowing, and believing run deep. The correlation between vision and our formulation of abstract concepts of reality, truth, and rationality dates back to the ancient Greeks and persists in varying degrees to the present day.[38] Regardless of whether this connection is culturally/historically determined or more biologically based (and Jay, for example, has shown that there are certainly cases to be made for both), for current visualization researchers it is embraced as a foundational principle for their research.[39] In this sense, visualizations of phenomena become models capable of both defining and communicating knowledge about the world in a process that Ben Shneiderman calls both "external cognition" and "expanded intelligence."[40] Visualizations, in other words, help scientists think about the phenomena they study and reveal truths that would remain hidden in the data itself, at least according to the scientists who work in visualization research. For Shneiderman, diagrams, charts, maps, and any number of other visual displays are tools that represent our observations of the world and enable us to discover new forms of knowledge about it.

If the renaissance in data visualization has a modern birthplace, it is arguably the Human–Computer Interaction Lab at the University of Maryland, which Shneiderman founded in 1983. The lab focuses on developing new tools and forms for visualizing large-scale data sets.[41] In a series of textbooks for the field, Shneiderman and his colleagues stake an intellectual ground and a scientific importance for contemporary data visualization that even echo intellectual histories in the field of visual culture. *Readings in Information Visualization,* which bears the subtitle *Using Vision to Think,* begins:

> To understand something is called "seeing" it. We try to make our ideas "clear," to bring them into "focus," to "arrange" our thoughts. The ubiquity of visual metaphors in describing cognitive processes hints at a nexus of relationships between what we see and what we think. . . .
> The interweaving of interior mental action and external perception (and manipulation) is no accident. It is how we achieve expanded intelligence.[42]

For Shneiderman and others working in the field, there is a clear belief in the untroubled connection between seeing, thinking, investigating, and communicating. This degree of faith in visualization's capability to record and reveal the world not only echoes adherents of the documentary image but also pushed the medium to prominence at a critical historical junction in the contemporary transparency debates.

Outside of academia, this same faith in the connection between vision, visualization, and insight is further echoed by the person commonly thought of as the popular-culture guru of the field of data visualization: Edward Tufte. Trained as a political scientist, Tufte began looking at the power of data graphics when he cotaught a seminar on statistics to a group of journalists alongside the famed Princeton statistician and graphics pioneer John Tukey. After completing a manuscript on the subject, Tufte became frustrated when an academic press refused to print his book to the exacting standards he maintained the subject matter demanded: high-resolution graphics, archival paper, reader-friendly formatting. He took out a second mortgage on his home to self-publish the book, which became *The Visual Display of Quantitative Information* in 1983. Instantly a runaway success, the book is now hailed as one of the touchstones in the field of data visualization, and his follow-up texts *Beautiful Evidence* and others continue in the same vein. Tufte retired from Yale in 1999 and now tours the world giving one-day seminars on data visualization to sold-out crowds of academics, graphic designers, software developers, and product managers.[43]

Considering his presence both inside and outside the academy (in addition to the seminars Tufte gives, he has consulted with dozens of private companies), Tufte's influence on the contemporary field of data visualization is difficult to overstate. And Tufte, in turn, stakes the greatest social and intellectual claims for it. Indeed, the opening lines of *Beautiful Evidence* might almost be the governing ethos of the field as a whole:

> Evidence is evidence, whether words, numbers, images, diagrams, still or moving. The intellectual tasks remain constant regardless of the mode of evidence: to understand and to reason about the materials at hand, and to appraise their quality, relevance, and integrity. Science and art have in common intense seeing, the wide-eyed observing that generates empirical information. *Beautiful Evidence* is about how seeing turns into showing, how empirical observations turn into explanations and evidence.[44]

The connections he makes between empirical observation, explanation, and evidence demonstrate that through data visualization what might otherwise be simple facts or observations about the world are transformed into narratives that account for a particular mode of existence. Given the potential for such evidence to mislead, he calls the creation and consumption of evidence presentations a "moral act" in which presenter and audience are tasked to hold one another ethically, intellectually, and factually accountable. While such claims might easily be dismissed as mere hyperbole, Tufte provides evidence of the mortal importance of the particular form in which we choose to present information. In his essay "Cognitive Style of PowerPoint," Tufte analyzes Boeing's use of the popular presentation software to present its analysis of damage to the space shuttle *Columbia* during its launch in 2003. He convincingly argues that the default hierarchical format of the software program was partially responsible for NASA's inability to isolate the problem that eventually led to the shuttle's destruction upon reentry.[45] While many have derided PowerPoint for its facile display of information, few have argued that it could also be a matter of life and death.

Given Tufte's focus on the practical implications and limitations of the data vi-sualization, it is clearly no coincidence that he started his work in the field as a po-litical scientist teaching the craft of visualization to aspiring journalists. Whereas Shneiderman and other computer scientists champion data visualization's ability to produce new cognitive insights, Tufte's background seems to temper his enthu-siasm, forcing him to consider the uses and misuses of visualization in a wider social and political context. Beyond seeing data visualization as a tool for scien-tists to better understand their observations about the world, Tufte recognizes its impact on the realm of social and political discourse. Instead of straightforwardly representing "what" is in the world, he notes the extent to which data visualiza-tions are further called upon to narrate "why" a given situation exists and dictate "how" an audience should respond.

What further separates Tufte from the more classically trained scientists cur-rently working in the field is his emphasis on the decisive role that aesthetic form plays in this process. Significantly, he relies on photographic metaphors as mea-sures of the relative merit of a given technique or form. Visualizations are evalu-ated on their "resolution," which for Tufte refers not to the pixel density of a pho-tographic image, but rather to the amount of data contained within a given visual space. High-resolution graphics like the sparkline can present thousands of points of data in a space no bigger than the average printed word, while low-resolution formats like PowerPoint slides contain no more than a few dozen. Visualizations are further gauged on their "clarity," which refers not to photographic depth of field or focus but instead to the extent that essential information is brought to the fore and extraneous details are excluded. Entire chapters of his book are devoted to detailing the dangers of what he calls "chart-junk" and "PowerPoint Phluff"—extraneous formal features added to low-resolution graphics to hide their inad-equacies. Throughout his work, data visualization is for Tufte an aesthetic form in which every choice should be made with an eye toward maximizing the amount of relevant data that can be represented by a given visual feature.

Tufte's prominence in the field of visualization was recognized in late 2008 when he was called to Washington to use his data visualization skills to help his fellow citizens understand one of the greatest challenges it had faced in nearly cen-tury: the rapid collapse of many of the country's largest financial institutions. The person issuing the call was a young, newly elected president seeking to make good on many of his ambitious campaign promises: Barack Obama.

OBAMA'S OPEN GOVERNMENT

With his election to the White House in November of 2008, then president-elect Barack Obama immediately set to work fulfilling what had been his campaign's most amorphous and perhaps most compelling promise: change. Capitalizing on voter dissatisfaction over the protracted wars in Iraq and Afghanistan and increasing anxiety over the global economic downturn, Obama's promise of change stretched from the specific (closing Guantánamo within a year) to the general (bridging the polarized, winner-take-all politics of the previous eight years). Attempting to demonstrate his intention to make good on these promises, his transition team established the change.gov website to outline and track progress on many of his administration's nascent policies. Although some of these long-term ideas would prove to be, like most campaign promises, overly ambitious, his administration's use of digital technology as a channel of open communication was a change in and of itself. Signaling an immediate break from the perception of secrecy and dissimulation that had for many characterized George W. Bush's White House, the transition team's early utilization of web technology for public communication became a hallmark of Obama's approach to governing.

While this embrace of technology marked a clear policy departure from his predecessor, for both Obama and the left it was simply an extension of both his campaign and the eight years of progressive opposition that delivered him to the White House. His candidacy had been characterized by its utilization of the net-roots strategies of organizing and fund-raising that had become a pillar of progressive activism during the Bush era. From his move to post the president's traditional weekly radio address on YouTube to his very public desire to hold on to his Black-Berry while in office, Obama signaled early on that he intended to incorporate technology into every aspect of the new administration.[46] Just as George Bush had styled himself the first "CEO president," Obama clearly wanted to claim the title of the first online or digital president.[47]

Nowhere was the embrace of technology more apparent than in the new president's approach to government transparency. On January 21, 2009, on his first full day in office, Obama issued a presidential memorandum with the subject heading "Transparency and Open Government." Part of a move advocated by several government watchdog groups to embrace "openness on day one," the initiative outlined in the memo described his administration's approach to sharing information about the government and directed various individuals to coordinate an "Open Government Directive," which would revise standards for releasing information to the public.[48] Reminiscent of the idealism surrounding his campaign and echoing the heady optimism surrounding Web 2.0 initiatives that had launched nearly a decade earlier, the memo outlines three general principles to be followed by all federal agencies: (1) government should be transparent; (2) government should be participatory; and (3) government should be collaborative.

Along with moves to declassify documents and speed up the response to FOIA requests, these new policies were roundly applauded by the coalition of groups that made up what was by then becoming known as the transparency or "sunshine" movement.[49] Along with other moves by his administration over the next eight years, they placed the Obama presidency in line with a series of efforts by democracies around the world to reinvent or refocus government and democratic participation using digital tools and open-data frameworks. Eventually, these would include the Public Sector Information Directive in Europe in 2008, the Open Government Partnership in 2011, and the G8 Open Data Charter in 2013, among many others.[50] Many of these public open-data initiatives were justly criticized for further institutionalizing a neoliberal framework of privatizing public infrastructure (in the form of data) and normalizing competition as equivalent to efficiency.[51] However, the general ethos of releasing public data for the public good would eventually prove to be quite fragile once Obama left office and a new administration sought to alter the relationship between openness and government data.

The influence of entrepreneurial webspeak on the presidential memo turned out to be more than superficial. In the years that followed, the Obama administration instituted a range of new IT-driven policies and transparency initiatives—from websites and dashboards to blogs and data feeds—under the direction of the government's newly created chief information officer, Vivek Kundra.[52] Kundra had made a name for himself in the public sector by spearheading a number of open-government initiatives in Washington, D.C., under the reformist mayor Adrian Fenty, many of which would eventually come to characterize the federal government's approach to technological transparency. For example, one of Kundra's projects collected and published data, on a webpage called the "D.C. Data Catalog," that the city routinely collected.[53] It included access to everything from crime statistics and arrests to applications for building permits and city maintenance requests. Kundra realized that all of this data was both public information and potentially useful to citizens but that there was currently no way to connect the audience with its source. To address that problem, he borrowed what was quickly becoming a well-established model from the business world and held a contest that challenged people to come up with apps that would make the data more useful. The result of these efforts was a portal of tools and information that Kundra and the city hoped would be "a catalyst ensuring agencies operate as more responsive, better performing organizations."[54] Significantly, the page also included links and tools that would "allow users to create and share a variety of data visualizations" and informed each user that "you can create your own visualization using already uploaded datasets or slice and dice data the way no one has before."[55] The site was so successful that it eventually won the Harvard Kennedy School's Innovations in American Government Award in Urban Policy.[56]

After the 2008 election, Kundra was appointed to the transition team as a technology adviser, and part of his charge was to replicate the success he had had in Washington at the federal level. The result was the Open Government Initiative, which undertook a range of projects to push government transparency onto the web using a variety of tools. These included everything from sites like

USASpending.gov, a dashboard intended to track and streamline spending on government projects, to the OpenGov dashboard, a metachart tracking progress toward transparency across several dozen federal agencies. Among all of these initiatives, the two that best exemplify the government's current approach to transparency are the data.gov and recovery.gov projects that launched early in Obama's first term. Together, they illustrate the potentials and the pitfalls the government faces as it seeks to make itself more open to its citizens.

RECOVERY.GOV

Roughly one month after taking office, Obama signed into law the American Recovery and Reinvestment Act (known eventually as the "stimulus package" or the "recovery act"), which appropriated $787 billion of tax relief and government spending to counteract the effects of the ongoing global economic downturn. Two days later, on February 19, 2009, the website recovery.gov went live, welcoming visitors with the following message:

> Recovery.gov is a website that lets you, the taxpayer, figure out where the money from the American Recovery and Reinvestment Act is going. There are going to be a few different ways to search for information. The money is being distributed by Federal agencies, and soon you'll be able to see where it's going—to which states, to which congressional districts, even to which Federal contractors. As soon as we are able to, we'll display that information visually in maps, charts, and graphics.[57]

This initial page also featured a video of Obama explaining that the unprecedented size and scale of the recovery act demanded new methods of transparency and oversight to "root out waste, inefficiency and unnecessary spending."[58] Toward the bottom of the page was a simple bar chart breaking out the amounts dedicated to different spending projects. In spite of its rather limited initial offerings, the original home page for the site already contained all of the elements that would be essential to later versions. The site was a first step in what was intended to be a complete overhaul of the federal information infrastructure, a test case for his administration's transparency agenda.

After two years of updates, the site eventually became populated with a great deal of the information it initially promised. It became possible to search through hundreds of thousands of projects by size, geographic location, federal agency, or subcontractor. A number of different tools made this information available via charts, graphs, maps, and other visualization tools that allow one to analyze and interpret the data. The "video center" on the site listed over thirty clips (totaling just over three and a half hours of running time), each of which explained how to navigate the site and its overall mission. As an information source, recovery.gov represented an ambitious attempt to document and display

FIGURE 5.1. The initial recovery.gov home page.

the flow of federal revenue, dollar by dollar, from the Treasury into the multitude of projects it supported and the jobs it created—a virtual showcase of the recently reembraced Keynesian economic principles. But as an object of visual culture, the site was perhaps even more groundbreaking. In the economy of information it utilized to document the impact of the stimulus package, recovery.gov represented a clear faith in the ability of quantitative information to sufficiently represent reality, and thus signals the partial advent of a postphotographic form of the documentary impulse.

Placing a nonfiction, multimedia text like recovery.gov in the same conversation as documentary film is not without precedent. As Tom Gunning has convincingly argued, illustrated lectures and narrated slideshows by social reformers like Jacob Riis place early documentary still photography in the pre-evolution of later

FIGURE 5.2. Recovery.gov circa 2011, after a great deal of the data about the stimulus package had been generated and visualized.

forms of voyeuristic, observational ethnography.[59] In this formulation, a series of still images narrated by the speaker provides the same format (photographic evidence, timed delivery, narrative progression) that would later be united in the form of the sound documentary film. Similarly, the combination of evidence, argument, and political narrative that recovery.gov offered might be seen as a third iteration of the form—one in which all elements sit side by side rather than being delivered sequentially.

The purpose of the site, moreover, was one that places it squarely within the documentary tradition, or at least that part of the tradition populated by state-sponsored films that effortlessly (if overtly) conjoin civic edification with political persuasion. On an explicit level, the site's goal of openness and transparency might be read as providing information and issuing a call to action—both common items on the documentary agenda. Indeed, the ominous presence of a large red button that adorned the top of every page that asks visitors to "report waste, fraud and abuse" simultaneously notifies them that they can take action while assuring them that action is being taken. Beyond its stated aims, the site also narrates for visitors, in basic beginning-middle-end structure, the story of how the stimulus package moved from being a piece of legislation in Washington to a series of concrete projects carried out in the real world.

The metanarrative at work here delivers the implicit political message that such spending works, and works for "real Americans." Part of a series of massive government spending measures intended to safeguard the economy from slipping into a depression, the stimulus package took its place alongside the Troubled Asset Relief Program and the auto-industry bailouts, which together set aside nearly $2 trillion to address the state of the economy. Of these, the stimulus package was the only clear example of a classic Keynesian stimulus investment by the public sector. While the other programs may have prevented the collapse of such iconic names in American business as General Motors and Bank of America, the stimulus package stood apart in that it was designed to inject new capital into the market. Within the Keynesian model, this public spending will then cascade across the economy as people paid by the government spend their paychecks on other goods and services, which income is then spent on other goods and services, and so on. The site's invitation to "track the money" is thus an invitation to witness what economists call the multiplier effect in action. Recovery.gov thus seeks not only to persuade skeptical conservatives that the historic spending levels were effective and necessary, but also to reassure taxpayers that they, too, would be the beneficiaries of the government's largesse.

Contrary to the fervor of the debate that surrounded it at the time, the size and scale of both the stimulus package and the economic threat it was meant to address were not unprecedented. The similarities between the Great Depression of the 1930s and what was quickly dubbed the "great recession" were widely discussed, and many parallels were drawn between the policy responses of Obama and of Franklin Delano Roosevelt. Indeed, these policy parallels might also be suggestively extended to the media that both administrations created in support of them.[60] After all, in attempting to carry out the controversial resettlement of destitute farmers and migrant laborers, Roosevelt's Resettlement Administration (RA) sponsored what eventually would become the era's most iconic and influential media representations. In addition to the well-known and widely distributed photographic work of Dorothea Lange, Walker Evans, and Gordon Parks, the RA also produced two thirty-minute documentary films meant to educate the public: *The Plow That Broke the Plains* (1936) and *The River* (1938), both directed by Pare Lorentz.

The River, in particular, offers both formal and rhetorical features that make its comparison to recovery.gov particularly productive. The film focuses on the mismanagement of the Mississippi River watershed over the previous century as the nation pushed westward and documents the ecological and social destruction that resulted. Using a mix of statistics, maps, and images tied together with voiceover narration, the film argues for the need to control the river and restore the damage, both of which would involve the large public-works projects that have come to be associated with the New Deal. As Paul Arthur has argued, the film directly positions new forms of technology as the solution to bridging the existing conflict between man and nature but presents these solutions in a poetic, lyrical style that blunts the heavy-handed role government would necessarily play in carrying them out.[61] In rhetorical terms, while the film does contain sufficient logos-driven data to convince the audience of the size and scale of the problem (acres of farmland flooded, percentages of deforestation, tons of topsoil erosion, etc.), its primary appeals are the pathos-laden images of the destruction itself and of the people whose lives have been ruined.[62]

Returning to recovery.gov, we find a similar mix of elements with a decidedly different sense of proportion and emphasis. Whereas films like *The River* utilize data and statistics to support an overarching framework of photographic images, recovery.gov utilizes images to support what is otherwise intended as a data-delivery system. Rather than using a map to demonstrate the context and scale of the subject as *The River* does, the maps on recovery.gov instead become navigational tools through which specific data points may be accessed. While both texts seek to document specific flows and the impact they have on people's lives, for one this flow is the photogenic tempest of the nation's largest waterway, and for the other it is the flow of capital for the nation's largest fiscal outlay. In spite of historically similar origins and an overall shared political purpose, the two objects thus utilize radically different media to achieve their aims.

A great deal of this difference can be explained by the technological and social contexts in which they emerged. For the average audience member of the 1930s, moving images were the primary portal to the wider world. For the governments and institutions that sponsored them, documentary was seen as the best means of edifying and persuading the mass publics who gathered to see them. Like the hydroelectric dams that *The River* promoted, sound documentary films like it represented some of the most cutting-edge mass-media technologies of the time. Seventy years later, this slot has been filled by the Internet. While the information agnosticism of the Internet's design stipulates no difference between the types of data it carries—all packets are created equal—in the information economy of recovery.gov, data as data takes precedent over image as data. The site certainly does contain images both still and moving, but these clearly play a supporting role in relation to the data visualizations that are its main focus. The vast majority of the site's three and a half hours of video, for example, are dedicated to tutorials on using the site's data interface. Hence, most of the images the website presents are,

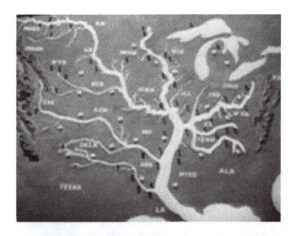

FIGURE 5.3. Data visualization in *The River* and on recovery.gov.

ironically, images of the website itself: screen captures, frame grabs, and so on. Both texts seek to "show" people what the government is doing to address their problems. But in earlier era, this meant photographic evidence. In the contemporary era, it means empirical evidence.

If data thus provides the core of the evidentiary claims that support the larger political argument the site levies, it also embodies the specific elements that form its basic narrative structure. Rather than the textual narration of the voiceover heard on *The River*'s recorded soundtrack (groundbreaking technology itself at the time), recovery.gov's temporal beginning, middle, and end are laid out on the horizontal time axis of the charts and graphs detailing the allotment of funds and the completion of projects. While this varies depending on the particular statistical lens one chooses to use, the site's focus on procedures like funding allotment and project tracking means that one nearly always encounters graphs trending in an upward direction as they move toward completion, subtly implying notions of uplift and progress as time moves forward. This impression is both reinforced and potentially predetermined by the framing the site's title provides. "Recovery" in a general sense refers to a process of moving from a diminished state to an improved state one had previously inhabited; but used in its noun form in this context, it also implies that this process is an object that might be purchased and put on display. While a given chart might be labeled "Funds Allocated by the Department of Education," the larger channel of transmission continually reminds its audience that what they are seeing is the recovery of the US economy from one of the greatest economic threats it has ever faced.

The site's "Lights-On Map" in particular demonstrates the combination of ethos, pathos, and logos at work in the broader rhetorical framework of recovery.gov as a documentary corpus. Created by Edward Tufte as part of his role on the advisory board for Obama's data and transparency initiatives, the piece was a map-based visualization covering the disbursement of funds allocated through the spending program. The map itself depicts the outline of the continental United States, with Alaska and Hawaii in an unscaled inset below the southwestern border. Dark, neon-blue land masses are set against a black background, evoking well-known satellite images of the earth at night. This impression is reinforced as a series of over 150,000 tungsten-yellow points begin to slowly illuminate different areas on the map, each corresponding to a disbursement of funds for one of the recovery projects. As they appear, the visualization strongly conveys the impression of city lights coming on, illuminating the darkness cast across nation. Two insets further illuminate the map's scope and purpose. The first is a text box over the northern border that describes what the map depicts; the second is a small chart that indicates spending dates and dollar totals for those dates, the total trending in an upward direction. The entire animation lasts less than a minute, even though final version spans four years and tracks $285 billion in funding.

Although largely respected within the design community, Tufte's map was greeted with muted enthusiasm in professional design circles.[63] Many thought he had overly simplified the information by making every point of light equal one award (regardless of its size) and missed a huge opportunity to make the graphic interactive with the mountain of data underlying it. It had, in other words, emphasized style over substance, committing in the process a number of the errors that Tufte himself so often decried in his many books and lectures on good design.

FIGURE 5.4. Edward Tufte's "Lights-On Map."

While not inaccurate, these critiques miss the larger rhetorical and historical work the map's aesthetic features reveal. The overt connection between spending money and bringing illumination to a space of darkness connects Keynesian economics with Enlightenment rationality and visual transparency. It further draws in metaphorical connections between the phrase "keeping the lights on," with its colloquial meaning of supporting struggling businesses, and the government's support of local economies and projects through the stimulus spending. As the animation progresses and the entire United States is illuminated, it draws on historical connections with the New Deal projects that *The River* had evangelized, including the Public Works Administration (which supported the construction of public infrastructure projects to generate electricity, among other things) and the Rural Electrification Administration (which supported extending electrical service to remote areas). Tufte's "Lights-On Map" might not have been the most data rich or interactive of the visualizations that appeared in the "Map Gallery" on recovery.gov, but it did aesthetically connect the policies behind the site's creation with the most ambitious efforts of Roosevelt's New Deal.

Beyond its evidentiary and narrative capacities, the overall site's transparency purposes bear out the extent of the government's faith in data's documentary capabilities. In addition to convincing people of the effectiveness of the program, the site also seeks to enlist the public's help in safeguarding these funds. By exposing the data to public scrutiny, any instances of "waste, fraud and abuse" will be highlighted and addressed. This implies a one-to-one correspondence between what's represented in the data and what's taking place in reality. This is a far more ambitious claim to transparency than the one advanced by the champions of film and photography. Certainly a film might succeed in uncovering some level of malfeasance or graft via hidden-camera techniques of the sort pioneered by filmmakers like Nick Broomfield. But a project like recovery.gov claims to have already supplied all the relevant information. Wrongdoing in one instance must be captured; in the other, it need only be uncovered. It is only through the sort of totalizing archive of documentation that such claims might be made.

One final note worth mentioning on the documentary efforts of recovery.gov relates to the appropriateness of selecting data visualization as medium to explore this particular subject. The 2008 financial crisis revealed the nature of post-Fordist capitalism for ordinary observers. Starting with the collapse of Lehman Brothers and continuing on through the collapse of the US housing market, the "credit crunch" faced by major US banks, and various other cascading factors and effects, what slowly became apparent was the completely intangible nature of money and wealth in the modern global economy. While money itself has always been an abstraction of sorts based on various forms of value created through labor and the exchange of goods and services, this abstraction took on an extreme form in the various exotic financial instruments that were partially blamed with causing the crisis in the first place. In an odd way, this level of abstraction found its perfect corollary in the data visualizations on recovery.gov. Though intended to document the process of repairing the damage to people's jobs and lives, quantitative visual media of this sort also ideally reflect the intangible nature of the information-based economy that caused the damage itself. While this data was intended to be a gateway to the multitude of real-world projects on which the money was spent, viewed cynically, it might appear to offer data as the solution itself.

DATA.GOV

Launched several months after recovery.gov, on May 21, 2009, data.gov was intended to be a clearinghouse for all of the government data sets already being collected by the different agencies that make up the federal government. Unlike the event-driven nature of recovery.gov, data.gov was intended to provide a permanent access point for the release of US government data, and hence remains online nearly a decade after its launch. Prior to its creation, the default approach to information sharing by federal agencies was typically to err on the side of secrecy, owing either to legitimate concerns over national security and individual privacy or to a more general fear that such information could lead to criticism or embarrassment by the agency that collected it.[64] As the White House's own blog described the situation: "For years, agencies have collected data in support of their particular missions. But before the ubiquitous use of technology, data often sat in filing cabinets and agency basements."[65] Even when agencies did release information, it often took the form of reports published in PDF format or as charts and graphs without any of the underlying data exposed, neither of which could be read by computers and other programs that might put the data to use. While this may have given agencies the sense that they were being open with their data, from a technological perspective it was no different than if they had been sitting in a basement or filing cabinet.

Responding to the president's memorandum on transparency, the release of the Open Government Directive in December of 2009 put in place official policies dictating how agencies should handle their existing data. Agencies were not only reminded of the three principles set out by the president ("transparency, participation and collaboration") but were also given a primer on the value of these principles, with details like: "Transparency promotes accountability by providing the public with information about what the Government is doing."[66] In addition to setting forth these general principles, the document sought to change the very culture of each agency, claiming:

> To increase accountability, promote informed participation by the public, and create economic opportunity, each agency shall take prompt steps to expand access to information by making it available online in open formats. With respect to information, the presumption shall be in favor of openness (to the extent permitted by law and subject to valid privacy, confidentiality, security, or other restrictions).[67]

The document also put in place specific deadlines each agency had to meet in order to be considered "in compliance" with the directive. These included milestones like placing three high-value data sets online within three months (which would then link back up to the data.gov site), as well as goals for targeting and publishing all of the data an agency collected. In order to hold the agencies accountable, the Open Government Initiative set up a dashboard measuring each agency on ten different benchmarks.[68]

The idea of creating a specific goal tied to a transparent progress report seems to have worked. When Data.gov launched prior to the directive, it featured just 47 different data sets. On its one-year anniversary, this number had grown to 250,028, and by the end of two years it was well over 379,000.[69] The site as it currently stands is a teeming mass of information, featuring lists of everything from the "Failed Bank List" published by the FDIC to the "Farmer's Market GEODATA" list put out by the Department of Agriculture. Although the updates in the site's various sections are admittedly uneven (the "Climate" section, for example, has had no updates since 2015, reflecting the priorities of the Trump administration EPA), many agencies are still publishing data on the site long after Obama has left office. Alongside all of the data sets are tools for viewing the data in different visualizations (charts, graphs, maps, timelines) as well as options for downloading the raw data in machine-readable formats (.csv and .xls). Beyond this, an "Applications" section featuring a number of ready-made tools and visualizations enables visitors to utilize the data and share their interpretations with one another, and a "Community" forum allows people to provide help to one another on using the site and feedback to the agencies for improvement.[70]

At first glance, it's difficult to see what use all of this data might be put to. In essence, this is part of data.gov's strategy. Rather than trying to anticipate what might be useful for people, the site is designed to give users complete access and allow them the freedom to create new views on the data by filtering and combining data sets to reveal insights. Responding to criticism that the site was too intimidating for average users (as one gender-biased review put it, it failed the "mom test"), visualization tools were added so that users could interact with the data right in their browsers and then save and share their visualizations with other users.[71] To further the aim of innovation and accessibility, agencies were required to create apps that visitors could use to navigate the data in meaningful ways, and several contests have been held by outside groups seeking to test the site's usefulness. Sunlight Labs, a project of the Sunlight Foundation, for example, held an "Apps for America" contest, which solicited projects from the community and awarded prizes for the best submissions.

Early on, the data.gov website had some high-profile successes—both in a general, public way and behind the scenes. Some apps, like Airport Status Service, enabled travelers to determine in real time whether a particular airport was experiencing delays, providing small conveniences to citizens. Others, like Hospital Report Card, enabled them to make important health-care-related decisions and push hospitals to improve their levels of care.[72] Behind the scenes, as agencies began standardizing and sharing data, some unexpected conveniences also emerged. The Department of Education (DOE) and the Internal Revenue Service (IRS), for example, found that by sharing data they could allow users to prepopulate the DOE's FAFSA (Free Application for Federal Student Aid) with information from their electronically filed income-tax returns.[73] Such things have the potential to be of enormous benefit to citizens simply by making data the government already has available to the public. All of these tools are designed to make the data that Foucault once deemed an essential tool of the nineteenth-century sovereign open to the twenty-first-century citizen.

For the most part, however, data.gov remains more of a potential benefit to citizens than an actual one. While it is still relatively new, the site has generated two well-placed criticisms that raise significant transparency issues. The first of these is the basic trade-off in the flexibility of completely unformatted data (what's referred to as wholesale data) and the more user-friendly but less flexible prepackaged data (retail data). Given varying levels of technological fluency, raw data will be unusable for most people. Put another way, it remains opaque rather than transparent. But retail data in the form of user-friendly charts and graphs flies in the face of the entire ethos of open-government initiatives by creating a layer of mediation between the information and its audience.[74] This conundrum between direct access to unfiltered material and legibility for the average user is the same one confronted by advocates of the Direct Cinema school in the United States. Eschewing manipulative intervention in favor of "fly on the wall" filmmaking without interviews, voiceover, or expository intertitles, figures like Frederick Wiseman nonetheless implicitly rely heavily on skillful editing and camerawork to shape their material into a comprehensible form. While the trade-off achieved a certain level of formal purity, it opened its adherents to charges of duplicity nonetheless.[75]

While proponents of transparency and open government tend to favor some combination of both methods (which data.gov offers), allowing a completely free and open interpretation of the data runs a significant risk of error and misinformation. Even supposedly user-friendly applications that make use of data.gov, such as datamasher.org, require a basic understanding of the way statistics work. The site, which won the Sunlight Labs "Apps for America" contest and is often held up as an example of the power that data.gov opens up, allows users to combine any two data sets, such as poverty levels and high school graduation rates, on a national map to reveal correlations between different facets of social life in the United States.[76] But the potential for misapplication of the tool's parameters and hence misinterpretation of the data is readily apparent. To use our previous examples, consider a map that correlates the location of local farmers' markets with failed banks, claiming to demonstrate some relationship between the two. While a numerical relationship between the data would be easy to map using the tools the app provides, it would be difficult to claim that any causal connection had been revealed. The simplicity with which these visualizations can be created, and their connections to supposed "facts" about the world, mask a complexity in the science underlying the meaningful information we always hope such things will provide.

Moreover, as these tools and visualizations migrate off of the data.gov website and into the blogs and forums where political discourse increasingly takes place, it becomes conceivable that numbers might be found to support any range of politically loaded interpretations. Kundra once claimed that he aspired to make data. gov so easy to use and share that it would play a role on blogs the way YouTube does, implicitly equating data evidence with moving-image evidence.[77] The open-source ethos Kundra brought to the project holds that any egregious misuse of information will be spotted and quickly corrected as it is in the Wikipedia model, where users act as editors and fact checkers for one another. But in the paranoid and relatively polarized atmosphere of online political discourse, it is equally plausible that such corrections will themselves be quickly dismissed. While the potential for misuse certainly isn't a reason to avoid providing the information in the first place, it is a cause to question the utopian aims to which the site aspires.

The second, equally serious criticism that data.gov faces deals with data that doesn't appear on the site. Many open-government proponents question the extent to which relatively useless data is giving the government the appearance of transparency even as the material most important for fostering greater accountability is left off of the site. This problem became an acute reality when the Trump administration took office. Transparency watchdogs were concerned about his apparent indifference to open-government initiatives more broadly and his outright hostility to specific factual data (like climate data) that conflicted with his policies. Sounding the alarm, many began archiving data before it disappeared from federal websites. While the widespread fears of "Trump's War on Data" did not come to pass in the manner that many feared, the anxiety created revealed the underlying fragility of the framework itself.[78] Data may not have been removed, but many of the sections within Data.gov haven't been updated since 2015, and there is nothing to prevent their future removal. As noted in the discussion of Second Life in the previous chapter, the Internet and digital platforms provide an effective means of making things instantly available to a widespread portion of the public, but the trade-off is what Wendy Chun refers to as an "enduring ephemerality."[79] Digital objects online can be available to everyone in one instant but gone forever in the next.

The inherent problem with gauging progress toward transparency is that most citizens and independent groups don't know the extent of the government's information holdings to begin with. As Ellen Miller of Sunlight Labs put it, "We don't like high-value data that involves [things like] wild horse counts. . . . We suspect they have data that would be of more interest to citizens."[80] As Aliya Sternstein points out, this might include information like which sections of private industry had been cited by the government for failing to meet public safety standards.[81] The paradox that all open-government initiatives quickly hit upon is that without complete and total transparency, it is difficult to gauge just how transparent the government is actually being. And yet, issues like national security and individual privacy do dictate the need for some "defense of secrecy," even if such concerns are often overblown.[82]

The Obama administration proved itself open to the criticism surroundings its new initiatives and capable of responding, when it decided to act, with at least a version of the "change" it promised. For example, after an extensive audit of open-data programs timed to coincide with the celebration of "Sunshine Week" in March of 2010, the National Security Archive's executive director, Tom Blanton, stated, "The Obama Administration deserves an 'A' for effort but an 'Incomplete' for results."[83] Sensitive to the criticism, the administration quickly announced on its blog a redoubling of its efforts, providing a "Tour of the Horizon," which reiterated that "transparency is one of the core principles of democracy."[84] The result was the launch of 2.0 versions for many of its sites that addressed the concerns brought by the administration's critics. But in an odd twist of irony, the technology-driven principles of transparency would be quickly put to the test by another open-government initiative, one operating outside of the Washington, D.C., beltway and headed not by an Obama appointee but instead by self-appointed activist-anarchist Julian Assange.

WIKILEAKS

The months following Sunshine Week 2010 would indeed be, as the White House blog predicted, "chock full of examples of concrete efforts—not lip service—to making open government happen."[85] However, these efforts were in large part the result of a collaboration between Julian Assange and Chelsea Manning to release several hundred thousand classified documents and other media from the US government's Secret Internet Protocol Router Network, or SIPRNet.[86] As if heeding Obama's call for transparency, collaboration, and participation, the government's open data was at once hit upon by a number of groups utilizing a range of new techniques to analyze, visualize, and make sense of the avalanche of information. As the ensuing drama and debate over WikiLeaks played out, a number of the groups that supported the Obama administration's initial open-government efforts once again stepped up to support the new poster child of transparency, in the process creating an uneasy juxtaposition between idealistic government bureaucrats on one hand and renegade anarchist whistle-blowers on the other. In spite of clear legal and procedural differences, however, the two groups shared a belief in the need for governmental transparency and a clear faith in technology as the best means of achieving this.

Any discussion of WikiLeaks and its role in the transparency movement confronts several potential problems. The first is the overidentification of the organization with its complicated and controversial founder, Julian Assange. Even his most ardent defenders and allies admit that Assange is an overbearing, attention-seeking figure prone to taking credit for the entire transparency movement. As accounts like Daniel Domscheit-Berg's *Inside WikiLeaks* make clear, both the organization and Assange himself benefited from a broad array of actors working toward the same ends.[87] The second problem is the overidentification of WikiLeaks with the series of massive data leaks that have come to characterize the era of big data itself. These include incidents across a broad spectrum, from Edward Snowden's shocking revelation that the National Security Agency was spying on US citizens to the incompetent release of personal credit information by Equifax in 2016. While WikiLeaks played an important role early on, as I will demonstrate here, not all leaks are politically, ethically, and materially equal, nor are they all the result of WikiLeaks . And finally, WikiLeaks itself has shifted radically from its early politically ambivalent anarchist stance to one more directly focused on shaping existing government structures through election influencing.[88] While I believe WikiLeaks deserves a great deal of credit for pushing the transparency debate in 2010, I would also caution against a simple celebration of the impact that it has had on democratic governance over the past decade.

Although WikiLeaks had been on the fringes of the activist and hacker communities for several years prior to 2010, its largest leaks had focused on fraud in private entities like the European banking giant Julius Baer and on political corruption in places like Kenya and Peru.[89] The organization had attained a reputation among hackers and transparency advocates for creating technology that would allow anyone to securely and anonymously upload large caches of previously secret data for publication on its servers. Its early successes also earned it the attention and appreciation of both the media and the nonprofit sectors, which recognized its positive disruptive potential for information freedom and social justice.

Throughout this early phase, WikiLeaks seems to have largely positioned itself as a basic conduit for information, publishing any and all contributions it deemed authentic and leaving the interpretation and investigation of the material up to journalists and activists. The site attempted to bring principles from the open-source software movement like community collaboration (memorably expressed in Linus Torvalds's "with enough eyes, all bugs are shallow") and open information exchange (Stewart Brand's equally memorable "information wants to be free") to the practice of whistle-blowing and investigative journalism. In April 2010, however, WikiLeaks began a series of high-profile leaks that would eventually earn it intense international legal and media scrutiny. The leaks, taken from classified US government databases, related to the wars in Iraq and Afghanistan and US diplomacy around the globe. Sensational as these releases were, the subsequent treatment of the source material itself reveals a great deal about the peril and promise of total informational freedom. While WikiLeaks and the trials of its controversial founder, Julian Assange, became a model for lofty goals like freedom of speech and governmental transparency in the age of the Internet, the migration of the leaked material across the media forced these ideals to confront reality.

COLLATERAL MURDER

The first of its high-profile leaks—the *Collateral Murder* video—was released by WikiLeaks on April 5, 2010, on the site's main page and on a connected site set up by WikiLeaks (http://www.collateralmurder.com). Both sites featured a set of videos drawn from an encrypted video file the site had received several months earlier captured by a camera mounted on the gunsight of an Apache helicopter in Iraq. The footage shows the helicopter shooting and killing eighteen people, including two Reuters journalists, and wounding two young children.[90] After vetting the footage to ensure its authenticity and sending people to Iraq to conduct interviews and notify relatives of the victims, WikiLeaks broke the footage into several versions. Among these was the original thirty-eight minutes of almost unedited material and a second, shorter version that excerpted relevant portions of the film and included both a prologue and an epilogue to the footage to provide context. In a possibly prescient move, this shorter version also solicited donations for the organization itself. Both films are accompanied by radio transmissions between the crew of the helicopter and soldiers on the ground. Although not exactly graphic, the footage is nonetheless horrifying to watch and troubling on a multitude of levels—not least because of the fact that, regardless of the context, it depicts violent death on a scale and in a manner that few Americans are ever forced to confront.

As an example of the WikiLeaks style of technological transparency, the *Collateral Murder* footage is something of an anomaly. In spite of the attention the film generated with its release in April of 2010, it seems to have been virtually forgotten in the wake of other subsequent high-profile leaks. These included the Iraq and Afghanistan war diaries and, of course, the now infamous "Cablegate" release, which exposed the dirty laundry of the State Department in a cache of diplomatic cables sent back and forth between various embassies and Washington. In all of the press surrounding the US diplomatic cables, the *Collateral Murder* film barely rates a mention except as part of the "other material" allegedly released by US Army Private Chelsea Manning through her access to the government's SIPRNet. Moreover, its status as video footage taken from a single incident sets it apart from the other WikiLeaks releases, which have almost exclusively consisted of document collections spanning broader time frames. And yet, there is a great deal this film and its treatment reveal, not only about the status of visual media and visible evidence in the information age, but also about the value of information more generally for citizens in the democracy where the political debates around WikiLeaks played out.

As was noted at the time, the shortened version of the film represented something of a departure for WikiLeaks. Until this point, WikiLeaks had contented itself with summarizing and contextualizing the information it released but left the material itself entirely unaltered. With *Collateral Murder*, however, the released version of the film approached something closer to an analysis of the raw footage, thereby partially editorializing a particular interpretation. The move was criticized on two fronts. One set of critics felt WikiLeaks hadn't simply contextualized and interpreted the shooting, but rather had misrepresented it entirely by leaving out certain mitigating details, including the possession of weapons by several members of the group fired upon by the helicopter.[91] A second set of critics felt WikiLeaks had overstepped its role as a self-branded leaker of information and had become something more akin to a news organization rather than a simple conduit connecting sources of information with the public. Even Steven Colbert, in a rare moment of seriousness, confronted Assange on the issue, stating: "You have edited this tape, and you have given it a title called *Collateral Murder*. That's not leaking, that's a pure editorial."[92] Assange, both here and elsewhere, justified the move by stating that part of WikiLeaks's promise to its sources included generating what he called "maximum political impact" with the information it released—an aim echoed in statements by other members of the organization.[93] While this usually meant partnering with journalists at major news organizations like the *New York*

Times, Der Spiegel, The Guardian, and others, in this case WikiLeaks took on the job itself. In doing so, WikiLeaks wasn't playing the role of editor or journalist, but rather the role of filmmaker. In doing so, what it created was a documentary.

The case for claiming *Collateral Murder* as a documentary is a fairly straightforward one. Whether we take John Grierson's oft-repeated if equally contested definition of documentary as "the creative treatment of actuality"[94] or Bill Nichols's more recent reworking that it "tell[s] stories with evidence and argument," documentary is generally accepted to consist, in varying degrees, of a creative or critical interpretation of events in the historical world, often with the intent of convincing viewers to accept this particular version of events as "true"—loaded as that last term may be. Regardless of the degree of nuance these definitions leave out, or the expansive domains into which documentary scholars and practitioners have recently pushed the canon, *Collateral Murder* clearly fits comfortably within these boundaries.

As evidence of this, consider the film's opening intertitles, which situate the footage historically and cast it within a particular critical frame. After a quote by George Orwell about the speciousness of political language,[95] we're given a brief synopsis of the event and informed that two of the men killed, Saeed Chmagh and Namir Noor-Eldeen, were Reuters news reporters, and we're shown images of both. We're then told that Reuters petitioned the US government to release the video under the Freedom of Information Act, and the ominous final title declares, "[T]his video has not been released . . . until now."

In documentary terms, this opening segment is clearly doing a great deal of work, or what we might call, after Jonathan Kahana, intelligence work.[96] It organizes the field of knowledge by orienting it both specifically—this particular event, these individuals—and generally, via the Orwell quote, as part of the larger struggle between truth and lies in political discourse. The pathos-laden background information on the reporters—that both were respected, talented, and, in Chmagh's case, survived by a wife and children—cements the event's status as a genuine tragedy. Indeed, this implicit emotional framework becomes explicit in the first still image we see of Chmagh, an image itself framed in the grief of the son who clutches it to his chest. Lest we miss the point, the film's very title has already rendered judgment on the event by declaring it not simply a tragedy but one involving murder—an overt allegation calling for a judicial response that would hold those responsible for the killing accountable for their actions.

Outside of the prologue and the title, the film goes on to annotate and edit its source material in a variety of ways, including cutting out nearly half of the video's original length, enlarging the image at specific points to highlight specific details, and supplementing its on-screen information with labels and arrows that identify certain figures and details in the frame even as the filmmakers, controversially, ignore others. These alterations and annotations further cement the piece's status as a documentary. Most of the footage left out of the shortened version of the film contains a second offensive by the helicopter crew some twenty minutes later in which three missiles are fired into a building believed to contain enemy fighters and weapons. As Raffi Katchadourian reported in the *New Yorker,* this second attack was arguably the bigger story, and a more open-and-shut violation of the "rules of engagement" followed by the US military.[97] While WikiLeaks may have passed up the opportunity for a second "smoking gun," the omission yields a more coherent "beginning, middle, and end" structure that works in the service of—to return to the definition of documentary—telling a story with evidence and argument. Simply put, this is what separates the document of the raw footage from the documentary nature of *Collateral Murder.*

These alterations of the footage also evince a need to clarify for the viewer details that would not otherwise be evident—a need that points to a curious indexical duality inherent in the source footage itself. In terms of rendering the event visible for its viewers, the source footage offers both too much and too little. On one hand, its original purpose has endowed the footage with a wealth of informational artifacts visible on-screen: the camera's position in space, its angle relative to the horizon, the exact center of the frame, the time of day, and so forth. And yet, for us as viewers, all of this information is relatively meaningless. On the other hand, the resolution of the image is far too poor to yield the relevant details that we care about in our attempts to understand what is happening. As evidence of the footage's insufficiency, we might consider the failure of both the pilot and the gunner to distinguish between a weapon and a camera, both of which were actually present on the scene. The disturbing nature of the footage comes not from the images themselves, but rather from our secondary knowledge about what they depict. Ironically, the maximum political impact that WikiLeaks sought comes almost entirely from the context surrounding the leaked material, of which the material itself is a fairly faint signifier.

If *Collateral Murder* is operating here as a documentary film, one with a clear political position and a set interpretation of events, Colbert was right to distinguish between what he called "leaking" and the "straight editorial" of the film. The editorial section has of course traditionally been that portion of the broadcast where the unreachable ideal of objectivity is momentarily cast aside and the sources of the text are able to voice their particular opinions. As a contrast, for example, more traditionally objective treatments of the material were offered by every major news outlet, from *Al Jazeera* to *Democracy Now!*'s Amy Goodman.[98]

What seems less obvious, however, is why this additional layer of mediation was needed at all. Journalism's traditional role within a liberal democracy is putting eyes and ears on the ground where citizens can't be to provide them with the information they will need to make informed choices. As Ulrich Keller has demonstrated, it was at the exact moment in the nineteenth century when war was no longer waged as public spectacle that the war correspondent was born.[99] If, as the rhetoric surrounding it claims, WikiLeaks can "transparently" connect the "source" of the information with the public, why has the middle man persisted?[100]

On one level, the answer is obvious and borne out readily enough in *Collateral Murder*. As the film demonstrates, the raw information itself is anything but readily intelligible. And if the average viewer can't understand what's going on in forty minutes of video footage without help, then she or he stands even less chance when the object of consideration is vastly more complex—say, a cache of ninety thousand documents. This is what makes the *Collateral Murder* video emblematic of the problems posed by the larger WikiLeaks project. Even with direct access to a visual recording of a single event, the need for interpretation, and hence mediation, is immediately apparent. In this sense, the public relies on journalists and filmmakers to process the information so that it can be made accessible to a general audience. By offering multiple versions, WikiLeaks is just fulfilling its desire to create what it calls "scientific journalism" by placing the original evidence alongside the analysis so that viewers might consult it to arrive at their own conclusions. But on another level, we might wonder if this leaves us any better off than we were before.

At the heart of what WikiLeaks offers is the belief that contained somewhere within organizational and institutional archives is information that can and should be made public, with the promise of justice, accountability, and, ultimately, truth on the other side. Steeped in the Enlightenment faith of reason's ability to deliver one to the truth and reminiscent of the eighteenth-century debate on the freedom of the press, it should come as no surprise that this promise is at once alluring and controversial.[101] Indeed, the term "scientific journalism" itself belies an uncritical faith that, simply provided with the evidence, all rational individuals might arrive at the same conclusions. Both sides of the transparency debate are in agreement on this connection between evidence and universal truth. Their disagreement stems from one side's belief that the public has a right to access the information and the other side's willingness to accept government secrecy as a valid trade-off for some other good—presumably, security. Whether the release of information is seen as liberating or threatening, however, the belief in the "truth" of the leaks persists.

This same logic even extends to the radical left, where Slavoj Žižek claimed, in what can only be read as a parody of Donald Rumsfeld: "The real disturbance [of WikiLeaks's release] was at the level of appearances: we can no longer pretend we don't know what everyone knows we know."[102] For Žižek, the truth contained in the disclosures isn't the mundane truth that this or that injustice occurred in the midst of war, which is already known even without direct evidence. Rather, Žižek's interest lies in what he refers to as the true face of the power wielded over us by the state: the power to commit murder or wage war and lie about it. This is a truth that is "made more shameful by being publicized."[103]

Indeed, the same sentiments were echoed with another high-profile release of secret materials from an unlocked archive—in this case, the Abu Ghraib images that emerged from the cameras of Charles Graner and Sabrina Harman in 2006. In relation to those images, Judith Butler argued that they, and the illicit manner in which they escaped, were a striking counterexample to the ongoing battle by the state to regulate the field of intelligibility for the public through its policies of embedded reporting.[104] For Butler, the problem of embedded journalism was the clear possibility that the journalist, the supposed objective party, would simply adopt the ideological viewpoint of the individual units or soldiers whose actions they were supposed to be covering.[105] And this possibility has been borne out by several recent analyses of the practice.[106] But alongside the high-profile role of reporters embedded within different units of the military there has also been a contingent of unilateral reporters covering the conflicts. Left on their own to cover the war as they see fit, these reporters are neither protected by the military nor subject to the subtle indoctrination this situation may create. Indeed, Chmagh and Noor-Eldeen were both operating unilaterally when they were killed, apparently attempting to cover events beyond the perspective, and hence the safety, of troop units moving through the area. Despite the long history of comparisons made between the camera and the gun, never has the connection between the two been so direct, so mistaken, and so tragic.

The irony, however, is that for all of the debate surrounding the role of both embedded and unilateral reporters covering wars, some of the most high-profile events have been the leaks themselves, from Abu Ghraib to WikiLeaks. If the recent war in Iraq is, indeed, the "most covered war in history,"[107] then the distinction owes as much to Julian Assange and Charles Graner as it does to journalists. And while *Collateral Murder* highlights the dangerous roles of both embedded and unilateral reporters on the ground, the next two WikiLeaks releases would bring a third type of journalist to prominence: the data journalist.

THE WAR LOGS: IRAQ AND AFGHANISTAN

In spite of the attention *Collateral Murder* received, WikiLeaks was apparently just getting started. Over the course of the next few months, the video would be joined by several subsequent leaks of classified material, including what became known as "the Afghan War Diary" and "the Iraq War Logs" on July 25 and October 22, 2010, respectively—collectively known as "the War Logs"—and a final release of US diplomatic cables on November 29, 2010. While each of these collections is unique and offers interesting points in its own right, I focus here on the War Logs collection for the contrasts it offers to both *Collateral Murder* and the US government's official transparency projects.[108]

Unlike its handling of *Collateral Murder*, WikiLeaks decided early on to partner with major news organizations for the War Logs, both to increase the impact of the releases and to outsource the work involved in identifying the relevant information contained in the masses of data they offered. Shortly after the release of *Collateral Murder*, WikiLeaks contacted both *The Guardian* and *Der Spiegel*, offering them a scoop on the next set of data in exchange for a simultaneous publication date. (*The Guardian* eventually brought in the *New York Times*.) All three news organizations spent about a month decoding the cryptic terminology of the reports, verifying the data against other sources, and determining what material would be of public interest. On the day that WikiLeaks released the full dataset online, all three news sources went public with the reporting they had prepared over the prior month.

Looking at the data, what stands out immediately is the extent to which a professional, third-party source is needed to interpret its contents and make it legible to a wider pubic. While the term "transparency" implies an unobstructed or unmediated connection between observer and subject, the War Logs demonstrate the necessity of several layers of mediation between the two. If the optical transparency of *Collateral Murder* required basic identifying labels and arrows to clarify its contents, then the War Logs would need an outright translation. Of the approximately ninety thousand documents it obtained, for example, the *New York Times* identified and published on its website a selection of the most relevant material (about two dozen documents, in total). A typical line from one of the documents reads:

> JCC REPORTS THAT IP REPORTED THE IED STRIKE ON CF CIV VIC MB 4265 9065. IP CLAIM THAT 1X LN AMBULANCE DRIVER WAS KILLED BY UNCONTROLLED SMALL ARMS FIRING BY THE CF CIV CONVOY AFTER THE IED STRIKE (SEE ASSOCIATIONS FOR DETAILS OF IED STRIKE). JCC NOTIFIED 4/101AA AND REQUESTED THAT THE CF PATROL AT THE SITE INVESTIGATE.[109]

Needless to say, this is hardly material that's clear to anyone outside of the military communication channels in which this particular terminology is used. In addition to the necessity of decoding the terse acronyms and obscure identifiers (e.g., MB 4265 9065), there is also the need to identify what of the underlying information is actually new. Both the Iraq and Afghanistan wars were extensively covered by news media around the world with whatever degree of bias. This daily reporting had covered many of the daily events that the War Logs recapitulated. Much of the leaked information was, in other words, old news. Given the scale of the leaks and the quantity of outside information against which it could be cross-checked, it is no surprise that all three news organizations turned to "database experts" in order to "mine the data."[110] Thus, whatever level of truth these reports contained, it would apparently need to be unearthed by specialized machinery.

What was unearthed, it turns out, was at once surprising and unsurprising, at least in terms of the news media and its relation to the official and unofficial information coming from the government. The unsurprising aspect of both leaks was the largely sensationless nature of the information they offered. Overall, the assessment of the three news organizations seems to be that the actual conflict was less promising than the official government assessment ("bleaker" in *The Guardian*'s terms, "more grim" in the *New York Times*'s). Aside from several specific revelations (e.g., that the Taliban likely used a surface-to-air missile rather than a rocket-propelled grenade in bringing down a US helicopter), the logs themselves reveal no smoking gun. In a post-Watergate, post–Iran-Contra era, this sort of official spin seemed to shock few. Ironically, the leak itself ultimately became the biggest story.

More surprising than the actual information contained in the War Logs was the coverage it received by the news organizations that had early access to it. Simply reading the headlines published by the three primary outlets, one would get the impression that the War Logs had unearthed the truth of combat itself. *The Guardian*'s website, for example, claims in headlines that the War Logs offer "the unvarnished picture" and "expose[s] the real war."[111] The *New York Times*'s coverage claims that the documents take us "Inside the Fog of War" and offer a "real-time history" of the conflict.[112] On closer inspection, however, all three of the outlets carefully qualified and circumscribed the information contained in the reports. *Der Spiegel*, for example, included an FAQ section on its website, detailing all of the qualifiers that should be taken into account when going through the documents (level of classification, source, etc.). The *New York Times*, the most circumspect of the three, notes:

> It is sometimes unclear whether a particular incident report is based on firsthand observation, on the account of an intelligence source regarded as reliable, on less trustworthy sources or on speculation by the writer. It is also not known what may be missing from the material, either because it is in a more restrictive category of classification or for some other reason.[113]

In short, the War Logs offer *another* account of the wars but not *the* account.

While the prospect of any definitive account is an impossibility, the cautious approach of the *New York Times* to the unofficially released official material is reflective of the uncertain nature of government transparency in general. So long as some information remains secret, a necessity argued for by all but the most radical advocates of transparency, there will always be the suspicion that true "truth" lies at an even deeper, more classified level. This sentiment echoes the concerns voiced by Ellen Miller of Sunlight Labs over data.gov cited earlier ("We suspect they have data that might be of more interest to citizens"). While any information may be potentially useful, the invisible information casts a shadow on the visible.

Outside of their narrative coverage of the War Logs' material, both *The Guardian* and the *New York Times* also put together interactive visualizations of the data they contained. Of these, *The Guardian*'s "IED" (short for "improvised explosive device") visualization stands out as an excellent example of visualizations gener-

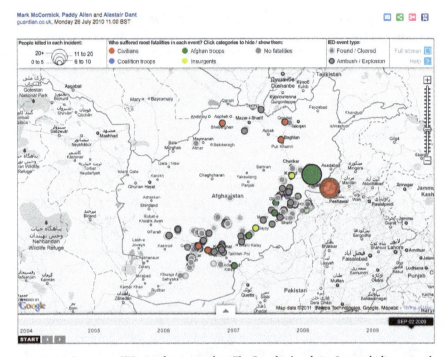

FIGURE 5.5. Interactive IED visualization tool on *The Guardian*'s website. In a multidimensional visualization like this, time and map location are overlaid with dots representing different attacks, with size representing the number of casualties and color representing the largest classification of those injured or killed (civilians, soldiers, etc.).

ated from the War Logs' data. Using data from the release of the Afghan War Diary, *The Guardian* sifted out reports of IED attacks in the country and used the embedded dates and GPS information to place them on a time-layered map of Afghanistan. Different targets and casualties were represented using both size and color codes (red for civilians, blue for coalition troops; the smallest dot for zero to five casualties, the largest for twenty or more, etc.). Viewed on a website, the resulting visualization can be played like a streaming video. Over the course of a few minutes, a map of Afghanistan is slowly dotted to reveal the number and location of IED attacks in a given time period.

In its own right, the visualization certainly opens up an interesting and otherwise unseen aspect of the conflict. The dots, starting slowly at first and building to a sustained crescendo, reveal in their concentrated location and apparent simultaneity both the contested regions of the war (primarily the beltway between Kabul and Kandahar) and the coordinated efforts of the Taliban insurgents. Even to a relatively attentive viewer of the news from the period, these aspects of the conflict become clear through the visualization in a way that they had not been previously.

Seen thus, a surprising quantity of information is condensed and made accessible in a relatively short time frame.

But beyond this initial impression, the insight such visualizations offer doesn't appear to penetrate very deeply. In other words, in spite of the quantity of information it provides (this would qualify in Tufte's terms as a "high resolution" graphic), such views of the event are hardly sufficient to understand the nature and meaning of the conflict itself outside of additional context. While one can certainly see that the number of IED attacks rises and falls at certain specific points, without additional sources of information it becomes difficult to ascertain the importance of any particular development. Seen through the lens of visualization, the War Logs reveal another account of these two wars, but it is very far indeed from revealing a definitive or total account. Even transparency, it seems, threatens to further obscure our view.

What remains most striking about the assertion of truth contained in the WikiLeaks material (both the video and the data) is the amazing capacity of the existing ideological frameworks to absorb this additional information as further evidence of their beliefs and move on. The left claimed that this material was evidence of the injustice of these wars, while the right claimed that it was a regrettable consequence of them. The extreme left saw evidence of a larger ideology of American empire, and the extreme right saw an instance of justice served. In spite of the millions of points of data added by the releases, the discursive framework around the events they record hardly changed at all.

Lest it seem that I'm sliding the debate over into the realm of an all too easy postmodern relativism, let me stress that I'm not claiming that all of these positions are factually or ethically equivalent, but rather that, despite their mutually irreconcilable positions, they remained unchanged in the face of new information. Nor am I claiming that transparency projects serve no purpose. As Micah Sifry points out, many such projects have achieved incremental improvements in government accountability.[114] But while the transparency movement may uncover cases of overt corruption and political manipulation, its potential for radical change against dominant ideology seems fairly limited. We certainly know more after WikiLeaks, but it seems we don't know any differently. In providing everyone direct access to the "truth," it seems, everyone's preexisting truths just become a little truer.

CONCLUSION

Writing on his blog "net critique," theorist, commentator, and sometimes agitator Geert Lovink published with Patrice Riemens what they called "Ten Theses on WikiLeaks."[115] Taken together, their observations amounted to an initial attempt to understand what WikiLeaks was by sidestepping the ongoing debate about whether or not WikiLeaks should be at all. The post contains a number of interesting points, concluding that WikiLeaks (and organizations like it) amount to pilot projects in what will be an ongoing process of greater information overload. Insightful in their own right, their observations stand out even further in this context for the insight they offer on the US government's transparency as well as the work of WikiLeaks. In their third thesis, for example, Lovink and Riemens point out that while WikiLeaks deserves credit for opening up US government archives, its efforts cannot be seen as ushering in the "age of global transparency" that many have claimed, given the extent to which other equally large players (China and Russia, to name only two) remain beyond the grasp of whistle-blowing prowess. The same might easily be said of Obama's open-government projects, which do a great deal to open up what is arguably already the world's most scrutinized government. Important though these transparency initiatives are, they hardly lay out a map that other governments will follow. Lovink and Riemens's point that WikiLeaks is a classic single-person organization (SPO), and hence rises and falls with the fortunes of its founder, might be translated with a few caveats to Obama himself. While data.gov survived into the Trump administration, nothing guarantees that a future administration will not summarily pull the plug in the name of security or cost-cutting or both.

While Assange and Obama are obviously not interchangeable, a closer inspection of both reveals the extent to which their efforts utilize surprisingly similar rhetoric to justify parallel projects that face identical challenges. Both official and unofficial transparency projects saw themselves as part of a larger open-source hacker ethos, working to provide raw material to the public to make whatever use of it they see fit. Both further neglect the essential role that the traditional apparatus of investigative journalism must play in achieving any real insight from the information provided.[116]

The move to data transparency hence necessitates the embrace (or the reembrace) of visualization techniques to render and make sense of it. This move from image-based media (film and television) to data-based media (online databases, the Internet) represents a regime change of sorts that has been in the making for much of the last two centuries. Thus, much as François Arago and Oliver Wendell Holmes were establishing photographic images as the gold standard of objective observation and documentation in the mid–nineteenth century, quantitative data and their visual display were busy opening up this other window on the world more recently. Over the last century, the two fields have continued to develop alongside one another. The widespread diffusion of camera technology promises a panoramic if not panoptoconic view of the world as developments in digital technology continue to generate an exponentially expanding quantity of data about it.

Even as digital technology has continued to erode a faith in photography's ability to unproblematically represent reality, a faith that was itself never entirely unquestioned, it has increased the role that data plays within it. Consider, for example, the extent to which debates about individual privacy have migrated from the fear that someone might optically witness and record one's physical actions to the fear that someone might access the information these actions inadvertently left behind: credit-card transactions, health records, and so on. If Lovink and Riemens are right in concluding that WikiLeaks is "a 'pilot' phase in an evolution towards a far more generalized culture of anarchic exposure, beyond the traditional politics of openness and transparency,"[117] then data visualization is the only means by which we might witness this evolution.

6

Post-Truth Politics

Conspiracy Media and the Specter of "Fake News"

After the midterm elections in 2014, as the Obama administration's second term wound down, the electoral calendar began looking ahead to the next election. Over the next two years one of the more remarkable presidential races in US history began to unfold, culminating on election night in November 2016 when something like a bomb went off in the American political system. Regardless of how one voted or which of the two major candidates one supported (if either), few anticipated that Donald Trump would win.[1] As the professional political class, the news media, and the general public all scrambled to understand "What Happened"—to borrow Hillary Clinton's phrase—multiple interpretations emerged to contextualize the implications of a Trump presidency and forecast its impact.[2] For many, it seemed to signify a radical break from the semiprogressive policies and cultural shifts around greater health-care coverage, environmental regulation, and inclusive policies and protections for the LGBTQ community. For others, it was a backlash against these same policies and a widespread reaction to the growing political clout of groups such as Black Lives Matter—confirmation that America's colonialist history of slavery and segregation was not as far in the past as many had thought.[3] For still others, it signified the emergence of a new form of ethno-economic nationalism, confirmation that the further entrenchment of economic inequality was taking root around questions of race and citizenship.[4] Regardless, virtually everyone agreed that something new had surfaced in American politics. With the surprise victory of President Trump to the White House, many felt that all of the accustomed frameworks and systems (journalism, democracy) were broken. Even "truth" was suspected to be in play.[5]

Of course, the full historical implications of this event will take many years to play out and touch on many different dimensions of American political and cultural life. Seen in the context of the preceding chapters, however, the Trump victory functions more like the fulfillment or perhaps the culmination of the larger trends around digital culture, independent media production, and mainstream politics that began in the period after 9/11. This concluding chapter will situate the 2016 election within the larger narrative threads that this book has explored. In what follows, I will examine how the Trump presidency and the shift in political culture that it signifies figure within an account of the evolution of nonfiction media as it evolved to incorporate and be subsumed by digital technology.

As the preceding chapters have demonstrated, post-9/11 political media and independent documentary both embraced digital technology in an attempt to find some novel form of expression capable of overcoming the discursive stalemate that set in with the ensuing war on terror. Be it the pervasive logic of computation in prominent documentary films (chapter 2) or the use of digital networks to facilitate film production and distribution (chapter 3), documentary filmmakers and their work seemed to be following a pattern that had historically played out several times before (chapter 1), even as this digital iteration led them to tools beyond the camera to produce nonindexical forms like video games, virtual environments, and data visualization (chapters 4 and 5). This period marked the moment when documentary became digital—not simply because of a material shift in production or distribution but because of a wider cultural shift in the world that these artists and activists were documenting. As early adopters who explored the limits and the capabilities of these newly emergent technologies, these media makers' digital documentaries prefigured and shaped some of the more emblematic qualities of digital culture, including a faith in data and the rapid, widespread adoption of social media.

Thus far the narrative has pursued two primary, interrelated threads. The first largely focused on the efforts of progressive activists on the left seeking to counter the apparent ascendancy of the conservative right. The second chronicled the extent to which digital technology eclipsed and enveloped virtually every other form of media. For some observers, these developments were necessarily connected, and President Obama's 2012 reelection seemed to signal a decisive victory on both fronts. It was seen by many of his supporters as proof that his first election hadn't been merely the result of an aberrant reaction to the 2008 financial crisis and was instead evidence of a larger shift toward a more progressive political environment on issues of race, foreign relations, and the role of the federal government. For others, the account of how he won was equally important evidence of the superiority of a newly decentralized and digitized approach to using data for campaigning and governing.[6] Among those digital utopians inclined to interpret the material base of digital technology as a progressive, liberating force of individual empowerment and democratizing potential, 2012 was a victory not just for the president but for technology itself.[7]

Of course, this was only half of the story, and the 2016 election demonstrates just how mistaken, or perhaps premature, these conclusions were. The ascendancy of digital technology owes as much to outside historical forces as it does to any particular political philosophy. And the marginalized role of the progressive left in the aftermath of 9/11 explains its prominent position in several of the preceding chapters. Political documentary has always been a tool of the opposition, whether it was opposing fascism (as it was in the '30s and '40s) or heavy-handed institutional narratives (as it was in the 1960s), championing the rise of AIDS activism in the face of official silence (the 1980s), or questioning the US government's response to an unfathomable terrorist attack after 9/11. In all of these moments, the opposition looked to new technology—be it sound, handheld cameras, video, or digital technology—as a way to revitalize the medium of documentary and counter the dominant discourse.

Rather than being an exclusive tool of the progressive left, however, digital technology and the documentary rhetoric of truth proved themselves, in 2016, to be open to any who cared to utilize them, regardless of their political leanings. Rather than a democratizing force capable of harnessing truth, digital technology proved to be a medium like any other. The demonstration of digital documentary as a medium capable of both truth *and* lies represents the point at which it became fully integrated within the documentary landscape. While it may have shattered the utopian hopes of some, in the context of this discussion Trump's victory completes and, in many ways, reinforces the processes that we have been following throughout. The events of 2016 are thus an important part of this story.

The problem, however, in confronting any narrative of the 2016 presidential election is that the narrative is one of multiple truths—not in a real sense, but as perceived in the minds of the electorate and expressed in the various media sources that attempted to inform and persuade it. That is, the "true" story is the presence of multiple stories, multiple levels of truth, and conflicting, contradictory versions of events. This fragmentation stems from two interlocking developments that connect to the larger ideas this book has been exploring. The first is the rise of conspiracy theory, and what I'll refer to as "conspiracy media," throughout the early 2000s. While conspiracy theorizing itself is nothing new, and not necessarily something negative, the post-9/11 period accelerated conspiracy thinking in a manner that illuminates an interesting dimension of post-9/11 political organizing, and nonfiction media more generally. This trend culminated in the election of a candidate who first gained political prominence by promoting a conspiracy theory about the legitimacy of the sitting president (Barack Obama) and his birth nation. While "birtherism" was roundly rejected in the mainstream media, Donald Trump used his notoriety as a reality television star to promote the fringe belief, thereby gaining entree to the political media and establishing a prominent brand of sensational notoriety that would eventually help him (in part) win the presidency. Conspiracy media within this environment benefited from and fostered greater legitimacy for themselves through the same cultural, political, and technological shifts—discussed in the preceding chapters—that benefited independent documentary. As independent media that levy strong truth claims, critique existing institutional and political power, and seek to inform and persuade their audiences, conspiracy media occupy an important (if idiosyncratic) part of the spectrum of documentary media more broadly.

The second thread this chapter will explore is the emergence in 2016 of a new species of media outlet, one that I will refer to as a "midstream media outlet." Using 2016 as a snapshot, I will explore the developments in the media landscape that produced one of the signature phrases of the 2016 election—"fake news"—and its origins in a more decentralized media landscape. Critiquing the mainstream news is, like conspiracy thinking and conspiracy media, nothing necessarily new. As chapter 3 demonstrated, the critique of Fox News was a pivotal issue in the hybrid experiments of left-wing groups like MoveOn, and Fox itself often positions its own approach as a remedy to what it calls the "biased" coverage on other networks. But in 2016, the topic became a central issue in the election. Indeed, critiquing the mainstream news became one of the signature moves of the midstream media outlets, a generation of news outlets that emerged online in the Obama years and that include names like *Vox, FiveThirtyEight, Breitbart,* and *The Intercept.* Positioned between mainstream news media and more extreme outlets on the fringe, these midsize outlets gained a prominent visibility in the 2016 election. As with conspiracy thinking, their emergence upends a great deal of the established thinking about the role of the news media and politics. Even in the context of digital media's disruptive rise across all segments of contemporary culture, news media and the nature of the industry that provides it have been radically impacted. Outlets that had existed for the better part of a century disappeared, giving way to powerful new competitors barely a few years old.

As one of the "discourses of sobriety" that Bill Nichols aligns with independent documentary film production, journalism represents a large portion of the landscape within nonfiction media more generally. Where chapter 3 was focused on questions of documentary film production and the use of digital media as a critique of right-wing news sources, and chapter 5 discussed the use of data-journalism techniques, the focus in this concluding chapter is on journalism from a more industrial, structural perspective: the nature of the news as a business and the manner in which it articulates and fulfils a role for itself within society. The rise of the midstream media outlets, and the political polarization of mainstream media more generally, have produced a new definition of what "real news" looks like—one that's increasingly influenced by the role of independent political documentary.

Individually, these two information flows appear to be polar opposites. Conspiracy theories are, almost by definition, ideas or interpretations held by a relatively limited number of people. Their proponents seek to expose what others have kept hidden from the public. News media, on other hand, at least nominally work only from what can be demonstrably proven. Journalists seek to establish a shared basis of information in the public interest, always in search of the elusive goal of "objectivity." Where the more extreme examples of conspiracy theory might be dismissed as the mistaken fantasies of lone individuals, professional journalists aim for and usually receive a wide degree of public trust. But despite the radically different positionality of these two forms of information (one on the fringe the other in the center), they have a remarkable amount in common. For example, both conspiracy theorists and news journalists regard themselves for the most part as public servants, or at least as working in the public interest. Both also regard their work as a necessary check on the state or other powerful institutional forces in public life and seek a wide audience for the information that they provide. Finally, both to some extent regard themselves as the arbiters or at least sources of the truth about events and institutions in the world. While conspiracy media may not rise to the level of consensus media, or enjoy the same level of trust, its emergence within this same time frame illustrates and responds to the same dissatisfaction with the consensus provided by mainstream media. The formal structures in which these shifts play out—newly deployed digital tools and platforms—are the very technologies that this book has been exploring thus far.

The fates of these information sources are tied up in one another even as they reflect differing movements throughout the post-9/11 period. The fringe began moving to the center and the center began working, in some way, more toward the fringe. The space between the fringe and the center is in many ways the exact space occupied by independent documentary media, sporadically supported on one side by the larger media industry (e.g., Errol Morris and Sony Pictures) and in conversation with fringe individuals and outlaws on the other (e.g., Julian Assange and Edward Snowden). It is difficult to overstate the extent to which I do not mean these terms as some sort of judgment on the quality of either form of information. My goal here is to demonstrate the way that their mutual movement was a result of the particular political climate playing out in a continuously evolving media landscape shaped and reshaped by emergent technology.

CONSPIRACY MEDIA: A THOUSAND THEORIES BLOOM

Given the tendency that Donald Trump the candidate demonstrated for associating himself with various conspiracy theories and figures like Alex Jones who propagate them, it should perhaps come as no surprise that his election generated an immediate explosion of conspiracy theorizing about the forces that put him there.[8] As news emerged that Macedonian teenagers were fabricating news stories to generate web traffic, that a group in Russia had run political ads on social media sites promoting various fringe beliefs, that Cambridge Analytica (Trump's campaign consultant) had dubiously obtained access to profile data about millions of Facebook users, various theories began circulating that Trump's victory was the product of a conspiracy of unseen forces.[9] A full accounting of either the theories that Trump himself has flirted with or those that have circled around him is well beyond the scope and purpose of this chapter, but their sheer proliferation across the political spectrum is enough to demonstrate that conspiracy thinking had come to dominate a significant portion of the political conversation. But as with so many features of the 2016 election, the prominent role conspiracy thinking played was less a new development than the culmination of trends that had been in place since 9/11.

Conspiracy theories have long been an object of fascination both inside and outside of the circles who subscribe to them. Scholars working in fields as diverse as political theory, philosophy, cognitive psychology, sociology, and cultural studies have all considered different aspects of the phenomenon.[10] Alongside this academic work exists a widespread market for popular catalogs and encyclopedias that describe the different conspiracy theories that exist—a depth and breadth of interest best exemplified by the publication in 2008 of *Conspiracy Theory for Dummies*.[11] Despite, or perhaps because of, this widespread interest, the topic is one of fierce debate. The ability of individuals to formulate and hold beliefs that run counter to verifiable evidence or group consensus, the epistemological foundations and origins of such beliefs, and the relative threat to the existing status quo—or the promise that such beliefs pose for it—has led to multiple divergent interpretations. Theories about conspiracy theories seem to multiply and diverge at a pace equal to the theories themselves.[12]

There are, however, several points from these debates that can help orient our understanding of the prominent role that conspiracy theories played in 2016. Much of the interest in philosophy connects back to Karl Popper's contention that conspiratorial thinking runs afoul of the unintended-consequences fallacy.[13] That is, conspiracy theorists mistakenly assume that political and historical events are always the outcome of conscious choices. Major events that seem to have no obvious explanation or motivation must therefore be the work of powerful unseen forces. As Popper points out, however, outcomes often run counter to the expectations of those who orchestrate their causes. While Popper's argument is cogent, and well taken among scholars who work in this area, it nonetheless fails to account for the broad variety of circumstances that seem to produce conspiracy theories. As Charles Pigden and others have pointed out, many such events are the result of intentional actions (e.g., assassinations, bombings, etc.), but the larger motivations or causal connections remain unknown or unspecified, and thus leave open a space for conspiracy theories to flourish.[14] This thread within analytic philosophy has produced a helpful framework for categorizing and evaluating the epistemic basis for different theories and the larger logical conclusions they invite. As David Coady demonstrates, conspiracy theories can be logically justified or unjustified and factually correct or incorrect—an important distinction given the political manner in which the term "conspiracy theory" is often applied.[15] As Coady and others are quick to point out, not all conspiracy theories prove to be incorrect (Watergate and the Iran/Contra affair, for example), and many that remain neither proven nor disproven may still be justified given other evidence (e.g., theories that the Bush administration intentionally fabricated evidence of weapons of mass destruction as a pretext for invading Iraq).

While this view offers a useful corrective to the traditional tendency to dismiss all conspiracy theorists as "kooks" or "weirdos," the logical and political implications of conspiracy theorizing must nonetheless be approached with care. In a foundational article from 1999, Brian Keeley claimed that conspiratorial thinking may be typically justified, and its theories are often correct, but that admitting the logical ground on which some theories are based opens the door to a corrosive chain of conclusions that make it difficult to believe anything.[16] Concluding that the government has the capacity to plan, execute, and then cover up massive events like the terrorist attacks of September 11, the Holocaust, the AIDS virus, and so forth is to imagine a set of capabilities so vast that it becomes difficult to stake out a ground beyond its control. How are we, as citizens, capable of countering this level of power? This further places these types of theories in an epistemological bind: evidence that counters the theory can simply be dismissed as further evidence of the cover-up, evidence that supports it must have somehow escaped the vast powers of the conspirators, but evaluating the difference remains a fraught endeavor. Finding a logical basis to support any beliefs or perceptions in this environment, much less resist the level of control and manipulation this presupposes, places one in an untenable position.

While such frameworks provide a logical basis on which to individually evaluate the vast and expanding universe of conspiracy theory, they do not account for why they have become so prevalent, nor can they demonstrate how any particular theory fits within a larger political or cultural framework. These questions have been taken up in the social sciences and humanities, starting most infamously with the historian Richard Hofstadter's polemical essay (and later book) that first appeared in the pages of *Harper's* in 1964: "The Paranoid Style in American Politics." Hofstadter's use of the term "paranoid" as well as the psychoanalytically influenced style of his own midcentury writing created something of a backlash by later scholars, who criticized his attempts at "diagnosing" and pathologizing those who engaged in conspiracy theories.[17] It did, however, offer enough of a history of conspiracy theories as well as an account of their narrative form and rhetorical style to inspire a broad body of work in response. Mark Fenster's more recent and more comprehensive analysis in *Conspiracy Theories: Secrecy and Power in American Politics* positions conspiracy theorizing as an outgrowth of both a well-justified cynicism toward power and an enjoyable form of interpretive, narrative analysis by the population.[18] The populist thread of many conspiracy theories can veer into fanatical and problematic territory, but it nonetheless remains a foundational element in American politics rather than a fringe reaction (as Hofstadter had positioned it). In a similar vein, Fredric Jameson has examined the emergence and presence of conspiracy logics in other mainstream cultural texts including films and television shows as a symptom of (and attempt to grapple with) the pervasive but invisible cage of late capitalist totality.[19] The tendency to question the official narrative accounts provided by the government and other institutional powers was clearly a well-established force within American culture (popular and marginal, fictional and nonfictional) long before the 2016 election. It is also, of course, a tendency shared by many independent political documentary films, which have a long tradition of challenging official narratives and providing alternative points of view.

What remains less clear is the role that media play within the formulation and spread of conspiracy theory—a class of media that, as I mentioned earlier, I will to refer to as "conspiracy media." Conspiracy media comprise the various forms of evidentiary media (including government reports, documents, audio and video recordings, photographs, and other items) that connect the cultural and political impulses behind conspiracy theories with the logical frameworks in which alternative narratives are assembled and tested. They provide a nucleus around which different conspiracy communities coalesce, and function as key points of exchange between the official or consensus understanding of an event and the alternative accounts that conspiracy theorists put forth. The legitimacy of conspiracy media (authentic or fabricated, relevant or extraneous, transparent or obscure) forms the ground on which the competing realities they represent are weighed and tested. And perhaps most importantly, their presence testifies to the absent or unseen forces that by definition separate the conspiracy theory from the official or consensus narrative. Looking at these conspiracy media—what forms they take, where they originate, what they respond to, how they are interpreted—reveals the changing nature of media and the particular media environments that foster them at a given historical moment.

The essential role of conspiracy media for conspiracy theory can perhaps best be exemplified by the archetypical conspiracy case of the Kennedy assassination and its relationship to what is known as the Zapruder film.[20] Abraham Zapruder was a Dallas businessman who happened to have brought his new 8mm camera with him to Dealey Plaza on the day of the assassination and was filming the presidential motorcade when the shooting occurred. His footage provided a seem-ingly essential visual record for the investigation, and worked in a broader way to cement the assassination within the public imagination and the expanding visual culture of the time. Unfortunately, as a piece of visible evidence, the footage left a good deal open to interpretation. Within Kennedy-assassination theories the Zapruder footage forms a foundational ground from which nearly every narra-tive account of the assassination works, and this includes the official report the Warren Commission produced. By capturing the moment of the assassination, this piece of footage should theoretically provide a degree of clarity or consensus around what took place, but it has had an opposite effect. Everyone believes that this piece of moving-image media is evidence, but "Evidence of what?" remains the question. As Stella Bruzzi convincingly argues, its meaning is ambiguous at best.[21] Without the Zapruder footage, and indeed without the myriad history of its ownership—who had it and when, who was allowed to see it, what they may or may not have done with it while they had it—assassination conspiracy would still undoubtedly exist, but it is unlikely to have occupied the prominent status in American culture that it has for much of the past sixty years.

The role of the Zapruder film as conspiracy media in the Kennedy assassina-tion offers a number of points of correspondence with 9/11 conspiracy theory. As a germinal historical event for conspiracy thinking, the terrorist attacks on 9/11 are often characterized as the twenty-first-century equivalent of the JFK assassination. Both events were theorized as traumatic national events that signaled the end of a state of perceived innocence on the part of the American public, or American culture more broadly. Both also spawned official inquiries (the 9/11 Commission and the Warren Commission, respectively—the former of which, Fenster points out, was directly modeled on its predecessor).[22]

But if both events resemble one another in their cultural and historical sig-nificance and the official and unofficial responses that they generated, the media environments that surrounded them were drastically different. Media coverage, both amateur and official, of course exploded over the forty years that separated them. Instead of the "twenty-six seconds in Dallas" that Zapruder's film commit-ted to historical memory, events on September 11 played out over several hours. Rather than a single piece of amateur 8mm color film, many of the key moments in the 9/11 attacks were recorded live on television and through countless ama-teur recordings, particularly in lower Manhattan. Where the events in Dealey Plaza had fewer than a hundred eyewitnesses, viewers from around the globe spent most the day on September 11 in front of their televisions watching. To borrow Mary Anne Doane's well-known framework, we might observe that the JFK assassination as represented in the Zapruder film exists within the temporal framework of the catastrophe, whereas 9/11 unfolded in the space of the crisis.[23] Paradoxically, however, this quantity of evidence did not make it any easier to comprehend or arrive at consensus on the nature of either event. Rather than clarifying or generating consensus about 9/11, the pervasive ubiquity of media coverage and visible evidence did nothing to eliminate the spread of conspiracy theory around the event.

Taken together, these two events and their attendant media coverage demonstrate a few essential features of what I am referring to as "conspiracy media." The first dimension of conspiracy media that these examples reveal is their productive capacity to generate further media. If conspiracy theorizing is an attempt to assemble an alternative narrative account of events or institutional infrastructures within political, social, or historical contexts, then conspiracy media are in part the sources of the information and evidence they utilize to achieve this, and in part the eventual media they create to encapsulate and express these theories.[24] As the various histories of conspiracy theory demonstrate, conspiracy theorists have always produced different forms of media (books, films, radio and television broadcasts) to make the case for their theories. This was particularly true with 9/11 conspiracy theory as it emerged and evolved in the accelerated information environment of the Internet and wider participatory culture of Web 2.0.[25] The widely viewed documentary *Loose Change,* for example, was released online in 2005 and then reedited and rereleased in several different versions not only as its theories developed but also as the film garnered more attention and a larger production budget. Of course, almost any form of media (e.g., a popular film or television show) produces other media that comment upon it, which may in turn produce further media still. As scholars of participatory culture have demonstrated, fan communities have an incredibly productive capacity to expand the media universe of a given text.[26] But conspiracy communities are fundamentally different from fan communities.

This difference introduces the second quality of conspiracy media, which is the oppositional, antithetical relationship they have with one another. As competing interpretations of the existing evidence, conspiracy media discount or dismiss other accounts, including of course whatever official narrative or record about a given event exists. These aren't just differences of opinion; they are competing interpretations of reality, alternative "truths" put forward that claim to be the definitive account of significant historical events. If *The 9/11 Commission Report* is true, then *Loose Change* must be false, and vice versa. This extends further to the various competing factions within the community that produce conspiracy media. The adversarial nature of conspiracy media thus feeds into their tendency to obscure rather than clarify the events they describe, which in turn generates the need to create further responses that refute the others, and so on. Fueled by the general skepticism and suspicion that characterize the conspiratorial mind-set and enacting the iterative, interpretive impulse that this suspicion generates, conspiracy media are the metastatic output of the larger conspiracy culture that produces them.

The final quality of conspiracy media that we can see at work in both of our primary examples above is that conspiracy media, unlike media in other contexts, tend to obscure rather than reveal the events they represent on a wider cultural or social level. This is the logical result of the first two qualities of conspiracy media outlined above: that they proliferate and replicate, and that their various iterations tend to be contradictory or mutually exclusive. Put differently, one of the fundamental qualities of conspiracy media is that they confuse rather than clarify the subjects they represent. Like the spread of doubt about the existence of climate change, the introduction of competing narrative accounts into the discussion of 9/11 or the Kennedy assassination seems to produce a broader confusion around the nature of these historical events.[27] This seed of doubt seems to originate in the initial evidentiary media itself and regardless of its quantity. That is, like the Zapruder film, footage of the Twin Towers collapsing reveals one reality for a particular conspiracy community, and another for those outside of that community. To some extent, this greater quantity of evidence seems to have further destabilized consensus, allowing an ever greater proliferation of various accounts, both official and alternative, about what transpired and why. Although the degree of immediate versus delayed public suspicion differs between the two, both frequently stand out as among the most "popular" (that is, believed by the widest percentage of the population) conspiracy events.[28]

But if conspiracy media proliferate according to a perverse logic of their own and serve to further spread confusion, they contradict a certain logical supposition of conspiracy communities more generally: that they fundamentally lack critical information. When there is disagreement about some fundamental event or circumstance, it is logical to conclude that more information is needed. And where critical information is lacking, speculation will emerge. As Cass Sunstein and Adrian Vermeule argue in an influential if flawed article on conspiracy communities, many are the product of what they call "crippled epistemologies"— intellectual frameworks where certain people "know very few things and what they know is wrong."[29] In some ways this seems obvious. A greater quantity of evidentiary pieces (i.e., conspiracy media) will yield a greater number of potential combinations, and hence a greater number differing narrative frames in which to assemble them. Like cryptographic-key strength, an added number of variables or a longer key length makes finding the one "correct" combination more difficult among the increased number of incorrect competitors. In this sense, it is more accurate to claim that communities in the thrall of conspiracy theorizing know a great many things, even if what they collectively know is (by definition) mostly incorrect. Conspiracy communities do not suffer from a lack of information but instead from an abundance of it.

But the nature of this formulation should still trouble us. This simply isn't how information is (or at least was) supposed to work. Throughout the historical moment that witnessed 9/11 and during which the war on terror was taking shape, an alternative narrative for the role of media and information and its relationship to truth was also taking shape. The widespread ubiquity of digital media fostered an expansion of the forms of amateur media production. Originally grouped under the heading of Web 2.0, these technologies, operating under the banner of participatory media, user-generated content, and eventually just social media, were ushered in on a wave of cyberutopianism steeped in an Enlightenment, positivist faith in the power of information and collective participation to root out misinformation and distortion.[30] This ethos is expressed not only in the much-hyped business manifestos and public-relations messaging that fueled Silicon Valley, but also in the critical reflections of, to varying degrees, figures like Pierre Levy, Chris Kelty, Clay Shirky, and Yochai Benkler.[31] "With enough eyes," Linus Torvalds assures us, "all bugs are shallow." To be fair, there are a multitude of examples where collective effort produces substantive results (Linux and other open-source projects) and even reasonable informational accuracy (Wikipedia being the standard example cited by more-utopian scholars). The validity of these cases notwithstanding, however, conspiracy media offer an insistent and instructive counterexample.

If conspiracy media's abundance simply runs counter to otherwise valid models for participatory culture, so be it. However, it also seems to contradict the information/truth coupling of conspiracy theorists and their critics. As Sunstein and Vermeule contend, conspiracy theorists suffer from crippled epistemologies, which they define as "a sharply limited number of (relevant) informational sources." The cure, or perhaps the prosthesis, they offer for this condition is more information. In an unfortunate turn of phrase, one seemingly designed to stoke the nightmares of their conspiracy-minded subjects, the authors describe providing this cure as a type of "cognitive infiltration."[32] This is a process whereby better, more accurate information is made available in order to counter the misinformation that is feeding the conspiracy theories. While most conspiracy theorists would presumably resist the idea of being cognitively infiltrated, most would also agree that more information is needed for the truth to be known. Conspiracy media, as I've applied the term here, provides a peek into the conspiracy, but it also implies that the true nature of the event remains hidden from view, contained in information invisible to the public: the hidden images, classified documents, or other media that could narrate the whole story, or reveal the whole picture. This unseen evidence is the structuring absence of conspiracy, an absence that can be addressed only by more information. Thus, for both conspiracy theorists and their critics, the answer to the problem of misinformation is more information.

Returning to Donald Trump and the wave of conspiracy media that accompanied his rise to power (both the conspiracies he put forth and those that his critics circulated about his election), we can see that their emergence is the fulfilment of a longer trend. Conspiracy theories and the production of conspiracy media were already on the rise in the 1990s, well before the tragedy of 9/11. Oliver Stone's *JFK* in 1991 and the Mel Gibson–starring thriller *Conspiracy Theory* in 1997 adequately bookend a decade marked by the initiation of the various Clinton conspiracies that would follow the couple through to Hillary Clinton's unsuccessful candidacies in 2008 and 2016. The more general and ubiquitous emergence of digital media in the 2000s accelerated their spread. Given the paradoxical cycle that conspiracy media feeds into and off of, it is easy to see how these circumstances arose, although less so to forecast how they might subside. In a functioning democratic society, the traditional mediator between a skeptical public and its government is the journalist—an independent, professional watchdog tasked with holding the government accountable and informing the public. Unfortunately for anyone tuning in to find answers, however, the news was experiencing a crisis of its own, one complicated by the president's repeated charge that it was "fake."

MIDSTREAM MEDIA AND FAKE NEWS

With the rise of Donald Trump within Republican politics, fake news became a topic in the news. The evolution of its prolonged news cycle unfolded in several stages, not unlike the various stages of grief popularly associated with a traumatic loss of some sort. The concept first emerged (or reemerged) on November 3, 2016, just days before the election, when Craig Silverman and Lawrence Alexander reported the somewhat shocking news in *BuzzFeed* that Macedonian websites were publishing false stories about the various presidential candidates in order to generate web traffic and hence Google ad revenue for their creators.[33] Trump himself then used the term in December of that year to refute claims that he would remain on as executive producer of *Celebrity Apprentice* during his presidency. In the first year of his presidency, he went on to use the phrase "fake news" in over 150 different tweets, although, as Steve Coll points out, the designation unsurprisingly has very little relationship to the veracity of the news story to which he attached it.[34] This, along with Kellyanne Conway's use of the phrase "alternative facts," ushered in a wave of scandalized outrage and denunciation, before moving on to a further chapter of grief and mourning. This stage was marked by ponderous hand-wringing over the end of truth as a category of human knowledge, and was of course accompanied by a related process of pointing fingers and assigning blame.[35] As this "new normal" settled in, we entered the last stage, acceptance, which was characterized by a repeated assertion that "fake news" is old news, and has a long history that dates back, in some formulations, to Plato's *Republic,* and at least as far back as the widespread emergence of print news in the nineteenth century.[36] As long as there has been nonfiction media, it seems, there has been a means of fabricating this nonfiction as well as a degree of concern over our collective ability to decipher the difference.

Fake news may indeed be very old, but it can nonetheless point us to what's new in the new media landscape in which it is has most recently taken root. Within the debates over the role of "fake news," stories also began circulating about various tools and software programs capable of easily manipulating live video and audio footage. Dubbed the "next frontier in fake news," tools such as Retiming and Lyrebird seem to have replicated the crisis of confidence that still photography experienced with the advent of programs like Photoshop several decades ago.[37] While there is undoubtedly a technological component of fake news, most of the examples discussed above (from the Macedonian stories on through those labeled by Trump) were relatively low-tech, pointing to the primarily cultural component of the problem. However misplaced, such anxieties do reveal a surprising degree of ongoing faith in the power of optical media to transparently and objectively record events before the lens of a camera. Simply put, the possibility of faked footage implies the possibility of real or objective footage from which it deviates, a potential long debunked by scholars of the documentary image. Moreover, the emphasis on the truth or falsity of visible evidence itself overlooks the extent to which even unaltered or "raw" footage arrives heavily mediated by virtue of the source that produced it and its motivations for doing so. These factors may enable tenuous distinctions between categories like news and advertising, documentary and propaganda, but often remain invisible under the transparent glare of moving-image evidence.

Part of what the fear over fake news activates is a deeper fear of or mistrust in the influence of media to direct public opinion. As students of media studies will immediately recognize, however, these fears are long-standing, and in fact historically synonymous with the emergence of media itself. Indeed, the rise of mass broadcast technology in the early twentieth century coincided with deep fears about its use to promote what was then being newly referred to as propaganda. As Mark Crispin Miller points out, the term "propaganda" etymologically takes on its current political and social connotations in the aftermath of World War I, when it was revealed that public opinion on the Prussian aggression of the so-called Huns leading up to and during the war was a result of specific interventions by the US and British information offices. In essence, people felt they had been brainwashed into the war by the government. By World War II, however, this fear had been redirected toward the German use of media production as part of what was increasingly characterized as a technological "war machine."

One of the supreme examples that illustrates this political polarization can be found in Frank Capra's 1940 film series *Why We Fight*. This series of information films was intended to inform soldiers, and eventually the general public, about the nature of the Axis war machine. While Capra made sure to include plenty of images of marching soldiers and powerful munitions, one of the primary tools that formed the enemy arsenal in his depiction was its use of broadcast technology to spread misinformation. This fear of Nazi media found ready purchase because a widespread popular perception had emerged over the prior two decades that broadcast media such as radio and multimedia forms like film (which combined sound and image) were capable of manipulating the public and brainwashing them through the use of invisible radio waves and provocative, agitational moving images. This popular paranoia functioned as a mode of what Charles Acland refers to as vernacular critique.[38] But in Capra's conception, this fear around the power of technology was reshaped to focus on a fear of what political ends it was applied to. Its ability to manipulate the public wasn't disputed. What separated lies from truth wasn't to be found in the medium itself but instead in the politics that drove it. Like the Soviet montage theorists Dziga Vertov and Sergei Eisenstein, the powerful impact of media on the audience was a feature, not a bug. What separated good media from mere propaganda was a question of one's political position, and the designation thus became a label that was used to differentiate the enemy's use of the media from one's own.[39]

These traditional nonfiction categories (news, propaganda, documentary) have become increasingly complicated in the flattened digital-media landscape that has emerged online over the past two decades, reaching an acute crisis point in the fake-news debates of 2016–17. Rather than an alteration in the materiality of moving-image technology from analog to digital technology, this crisis originates in the uncertainty generated by the shifting political economy of the media market enabled by digital technology. Two particular manifestations of this evolution can help illuminate what is ultimately a broader shift. The first is the general democratization and widespread adoption of the tools of news manipulation—a practice commonly referred to as "public relations." The second is the emergence of independent, midsize news outlets like Glenn Greenwald's *The Intercept* or Nate Silver's datacentric *FiveThirtyEight*, among others. While these outlets offer a useful corrective to the corporate-dominated univocality of the "mainstream" news, their emergence over the last several years overturns several of the commonly held critiques of the mainstream media that have guided media studies scholars for the past several decades. They also, in part, feed the adversarial environment in which the suspicions around "fake news" were able to take hold.

News and documentary film both traditionally occupy adjacent positions on one part of the nonfiction landscape—a landscape that, as scholars of documentary have long pointed out, stretches across an amazingly wide variety of forms, contexts, and creators. It includes the types of amateur media production with limited distribution that we more commonly refer to as home movies and family-photo albums, as well as the vast swath of zero-degree video recording used for simple archival and broadcast purposes.[40] Here I'm referring to things like TED talks, university lectures, training videos, industrial films, and so forth—texts that are "true" simply because they offer an approximate record of an event. It stretches, further, to institutional contexts like surveillance videos and photos, mug shots, the ID photos on things like passports and driver's licenses—records that are "true" in an institutional regime of governmentality because the facsimile of people and events furthers and enables a level of control and regulation, as loaded and biased as these supposedly objective records might appear.

Within this spectrum of nonfiction material, the lines within and between these various forms of journalism and documentary and related fields like advertising and public relations are fuzzy.[41] In a sense, journalism, documentary, and advertising all attempt to use the media to alert the public to events or information in their interest. All of them are true in a loose way, but we divide them into different categories for good reason. The makers of a given product believe, to some extent, in the claims they make for it, even if the wider public may not be convinced, and an established set of laws and standards holds them to account for the claims they make. In the formulation of scholars like Roland Marchand, advertising is an attempt to persuade the public to take a specific action (buy this product), whereas news is (at least traditionally) intended to inform the public about a given set of issues that may have many different possible interpretations or observations.[42] Documentary, in most theorizations, sits somewhere in the middle, advocating for a specific position or course of action but not necessarily directly tied to a commercial outcome in the same way that advertising is.

Within this same landscape sits the field of public relations, or PR, as it is more commonly referred to. Public relations is the attempt to transform advertising into a form of news, or to steer the public discussion and perception of an industry or product toward a particular position or opinion. Like journalists, early pioneers in the field claimed that their work was performing a valuable public service: alerting individuals within the community to new products, services, or community developments that they might have otherwise missed. As Stuart Ewen describes in his monumental history of the field, "The rise of public relations is testimony to the ways that institutions of vested power, over the course of the twentieth century, have been compelled to justify and package their interests in terms of the common good"—even if this common good was not at first apparent to the public itself.[43] In orchestrating the alchemy that might transform advertising into news, PR inherently destabilizes already porous categories, thereby creating suspicion around the role that news plays for its audience.

The man considered by many to be the "one of the most influential pioneers of American public relations" was, of course, Edward Bernays.[44] Bernays, who passed away in 1995 at the age of 103, was a distant relative of Sigmund Freud—a relationship he seems to have implicitly put to good purpose in his own work. Beginning in the 1920s, he pioneered the practice of framing corporate goals and sales targets as events of public interest, finding connections between the two in order to garner free publicity for his clients. As the title of *Crystalizing Public Opinion* subtly implies, Bernays saw his own work as an instrumental corrective to the sort of intellectual labor and propaganda that Walter Lippmann had theorized was so essential and so corrosive to the working of a democracy in his own *Public Opinion*. Throughout his three books, Bernays's work comes across as an interesting mix of intellectual self-justification and crass self-promotion, the marriage of which was to be one of his singular, if dubious, achievements.[45] While PR consultants might have worked in the interests of their institutional clients, they nonetheless thought of themselves very differently from advertising agencies. As Bernays is quoted in the introduction to his landmark text, *Crystalizing Public Opinion,* public-relations men don't work in images (like their advertising counterparts) but instead "deal in reality."[46] This connection between image and reality, the attempt to interweave and mold both, is essential to understanding the connections between news, visual culture, and truth that fake news entails.

A comparable work written nearly a century later by the provocateur and inheritor of Bernays's mantle, Ryan Holiday, demonstrates the extent to which the manipulation of journalism and public perception has and has not changed in the intervening century. Holiday was a self-taught marketing consultant who worked with several high-profile if controversial clients including American Apparel and the author Tucker Max, eventually building a list of blue-chip corporate clients. In 2010, Holiday published *Trust Me, I'm Lying,* a tell-all about his methodology for manipulating the structure of the news industry. The book set off a minor controversy in media circles—an achievement he claims was intended to make the book self-exemplifying or meta-evidence of the methodology he utilized on behalf of his clients.[47] The book is thus one part media critique, one part how-to on media manipulation. Holiday's method is rather simple. He claims that the news industry is a deeply hierarchical structure where lower-tier blogs and social-media figures provide the content and fodder for publishers higher up the food chain in exchange for traffic and attention. The arrangement is informal and unwritten but, in Holiday's formulation, entirely predictable. Items that get attention at one or two lower-tier venues can be used to push higher-tier venues into coverage. For example, in the campaign for the Tucker Max movie *I Hope They Serve Beer in Hell* (Bob Gosse, 2009), Holiday purchased intentionally provocative billboard space around Los Angeles. He then defaced these ads, took photos of the vandalism, and e-mailed them to a local culture website as evidence of the controversy the movie was generating. Once the blog picked up this story and ran it, this coverage, along with the original photos, was then sent to higher-tier publishers, with a demand to know why they were ignoring what was becoming a movement to ban the film. And so on. Holiday in essence had pioneered, or at least claimed to pioneer, the sort of manipulative tactics that right-wing provocateurs like Milo Yiannopoulos and others would master. Put differently, Holiday had figured out how to use the type of trolling tactics that Whitney Phillips meticulously documents in *This Is Why We Can't Have Nice Things* in the service of promoting his commercial clients.[48]

While Holiday demonstrates an essential historical continuity with Bernays in his attempts to influence the media, he is also evidence of a radical difference in the media landscape of our current moment. Both Bernays and Holiday share a certain methodology: package the interests of a specific constituency (often a corporation or industry) in a way that makes it legible to the public—and more specifically, to journalists—as legitimate news within the public interest. But Bernays is clearly a man of institutions and the elite. He represented an industry, founded a council, and consulted with the institutional leaders at the highest levels of government, academia, and industry. Holiday, on the other hand, is far more of an individual, neoliberal player. His intended readership is not the intellectual elite, or even the individual of public opinion, but instead the individual who might be able to perform a type of armchair public-relations campaign for whatever pet project or product he or she might have to hand. Rather than laying out or attempting to establish an intellectual framework that might justify the role that public relations plays in American life as his predecessor had, Holiday instead offers us a confessional tell-all on the manner in which the blogosphere might be manipulated in order to achieve viral traffic and public notoriety.[49]

While it is tempting to dismiss Holiday's work for what it is (a straightforward attempt at self-promotion), we should also account for the acute critique that he offers of the economic structure of the blog-driven media landscape. In ruthless detail, Holiday delineates the metrics that drive blogs like *Huffington Post, Tech Crunch, Weblogs Inc., Gawker,* and countless others to underpay their writers on a per-story basis predicated on the traffic that those stories generate. These working conditions force writers to search for prepackaged, sensational content in order to churn out stories that achieve viral traffic levels. Since these same realities exist up and down the hierarchy of the media structure, even the largest, most reliable news sites depend in some way on the groundwork carried out by lower-tier, less reputable players in the food chain. These are the economic realities that drove Macedonian teenagers to create absurd news stories, Facebook to ignore the viral contagion across its network as people "liked" and shared the stories, *BuzzFeed* to cover the emergence and influence of completely fabricated news, and an endless stream of high-profile media outlets to pontificate on the ongoing presence of truth in our current moment. At each stage of the process, traffic and attention were the key metrics that drove the proliferation of content, until eventually traffic and attention became the content.

This may seem far afield from the discussion of documentary and digital media that has dominated the discussion thus far, but we should recall that the same forces in play here (greater access to the tools of content creation, suspicion of official accounts, political polarization) are what have driven both the alterations in the news landscape that are being described here and the widespread experimentation in documentary form that was described in preceding chapters. The economic realities of journalism are nothing new, but in the past these forces colluded to produce the norms and standards that the specter of fake news seems to threaten. In the United States the history of the news is driven by changing relationships between advertisers, journalists, and readers. As Marchand demonstrates, many of the early newspapers and magazines self-consciously crafted their content in order to assemble the audience that they thought would appeal to different advertising demographics.[50] Qualities like "truth" and "objectivity" became self-conscious ideals that were seen to appeal to a specific, well-heeled demographic that was desired by advertisers. The idealized model of journalism that emerged through the efforts of early newspaper titans like Joseph Pulitzer sought to maintain a strict boundary between the editorial positions of its publishers, the news it produced, and the advertisements it ran in support of the enterprise more generally.[51] For Pulitzer these distinctions were aimed at increasing circulation: creating a product that people would trust was a means of assembling an audience that could in turn be sold to advertisers. Along with the rise of the Associated Press style as the default tone for news coverage, an amorphous notion of truth and objectivity became the default ideal in American print journalism, even if there were many critics who pointed out how distant or impossible this ideal was. As broadcast media (first radio, then television) emerged, this same basic equation between content, funding, and audience was further elaborated through the mandate that this content should be in the public interest and regulated through broadcast licensing.[52]

This is not to say, of course, that news (in any of these forms) had achieved objectivity or truth. The equation on which it rested, paid for by advertising but produced in the public interest, was always an uneasy if not contradictory compromise. This inherent conflict has provided a nucleus for a great deal of media and communications research in the post–World War II period, beginning with early Frankfurt School work on the culture industry, and resurfacing in various arguments and key texts over the past fifty years. Consider, for example, Daniel Boorstin's much celebrated *The Image: A Guide to Pseudo-Events in America,* which analyzes the impact of visual culture on journalism as simultaneously cementing a notion of visual objectivity while playing to the basest, most sensational appetites of the audience in order to command their attention.[53] Or Edward Jay Epstein's equally central *News from Nowhere,* which vividly chronicled the production of network news coverage and the competing demands that it satisfied (advertisers, ratings, regulators, and, of course, the audience).[54] Other work, like *Manufacturing Consent,* by Edward Herman and Noam Chomsky, demonstrates that the illusion of objectivity invests the media with a powerful ideological capacity that overwhelmingly benefits the financial and political interests that shape it.[55] Further, the long-running concern over increased ownership concentration (the number of different media corporations that produce the news) also positions the business of media production against the quality of the information it provides.[56]

But even the most trenchant critiques of the news and its failure to achieve objectivity subtly reinforce the notion that this is the standard to which it aspires. Truth and the public interest are held throughout as the norm from which the media and journalists deviate. And, more importantly for my purposes here, it is the perceived failure to meet these ideals that became the basis of the charge of "fake news." We can see the contours of these competing forces if we consider the role of the news in covering the 2016 election. The phenomenon of Donald Trump as a candidate posed a conflict for news organizations: on one hand, his critics were predicting that his presidency would be a disaster. On the other hand, covering the disaster in progress produced record-breaking ratings for news organizations that had been under threat of consolidation and retrenchment for much of the period between 2000 and 2015. This conflict of interest was clearly expressed in a speech by the former CEO of CBS, Leslie Moonves, when he stated of the network's coverage of Trump's campaign, "It may not be good for America, but it's damn good for CBS."⁵⁷ This was also the same set of forces that had apparently driven the creation of the Macedonian news stories: outrageous information generated traffic, if not truth. Once again, the public interest appeared to be at odds with the business of journalism.

But the 2016 election cycle also demonstrated the increased influence of midstream media outlets. Sites such as *Vox, FiveThirtyEight, Breitbart, The Intercept, Quartz, Slate,* and *Axios* had positioned themselves midway between mainstream news and more-marginal independent bloggers and fringe news sources. While these were the most prominent, many more also appeared in specific vertical-content categories.⁵⁸ Many of them, in an effort to generate trust in readers and attract top-tier talent, are connected with or headlined by veteran journalists who once worked with older mainstream media outlets. Many employ techniques that are dependent upon or critical toward the practices of traditional mainstream journalists. And several of them are independently funded by wealthy individuals as ideological or political projects. Rather than producing consensus, these various outlets instead seem to respond to the polarization and dissensus that so characterizes the environment in which they emerged.

As an illustration of this, we can briefly compare two midstream outlets: *Breitbart* and *The Intercept.* On the surface, each resembles the other, and both fit the midstream model described above. *The Intercept* is largely backed by Pierre Omidyar, the billionaire founder of eBay, through an umbrella media project he started called First Look Media.⁵⁹ The founders of the site—Glen Greenwald, Laura Poitras, and Jeremy Scahill—all came from existing careers in media and journalism, listing among their credits *Salon, The Guardian, Democracy Now,* and the *New York Times,* although Poitras was best known as an independent documentary filmmaker. Similarly, *Breitbart* was founded by veteran blogger and *Fox News* commentator Andrew Breitbart and heavily funded by the Mercer Family

Foundation.[60] After Breitbart's death in 2012, Steve Bannon became executive chairman and worked on revamping and expanding both the website and the organization itself. Ideologically, the two sites could not be further apart. Under Bannon's leadership, *Breitbart* positioned itself as a platform for the far right and "alt-right," openly supporting extreme conservative political issues and candidates, including Donald Trump. *The Intercept*, on the other hand, is largely considered liberal or left-leaning, although it would balk at any political affiliation. Both sites position themselves as correctives to the failings of the mainstream news. For *The Intercept*, this takes the form of what it calls "adversarial journalism," pointed at the traditional sources of power, including older, established media like the *New York Times*.[61] For *Breitbart*, it means countering what it perceives to be the liberal bias in most mainstream media. What is striking about both is the extent to which their shared qualities (independent funding, a critical stance toward mainstream media and other sources of power, argument-driven information) are qualities that might equally describe a particular iteration of documentary film itself. Indeed, it is not a coincidence that figures like Scahill, Poitras, and even Bannon had, in other contexts, turned to documentary in order to accomplish what they now turned to journalism to achieve.

The evolution of these midstream players and their "adversarial" relationship with their competitors is the latest development in a longer chapter in the history of the news. Beginning in the 1990s, the rather stable landscape of mainstream journalism, consisting for decades of print media and the major broadcasts networks, was invaded by newcomers from cable, radio, and eventually the Internet. But rather than unifying mainstream media's homogeneity, each new voice positioned itself as an essential antidote to some key deficit in the existing sources of information. Thus, CNN was the answer to the limited amount of time allotted to news by broadcast networks, Fox was the answer to the liberal bias of CNN and others, MSNBC was a response to Fox's overt right-wing bias, bloggers were the answer to the class of professional journalists, and so on.[62] Each new entrant to the increasingly competitive news marketplace—itself newly marketized as a result of media consolidation and shifting revenue sources within older, established forms like newspapers and broadcast networks—thus joined the mainstream by critiquing its existing framework and positioning itself as a better instrument of truth and objectivity. This market-driven sense of distrust also came amid cogent critiques of mainstream media practices from figures like Jon Stewart, Steven Colbert, and even *The Onion*, attracting large audiences while delivering well-justified outlines of mainstream media's shortcomings.

This reaches something of a crescendo in the emergence of midstream media. All of them position themselves in some way as the antidote to what's wrong or what's lacking in mainstream media. *FiveThirtyEight* prides itself on its data-driven journalism, and the manner in which objective data might counter the affective, subjective impressions of reporters that plague traditional journalism. *Vox* is the explainer news source, giving the story behind the stories that other news outlets take for granted. *Breitbart* and *The Intercept* are both independently funded, hence not reliant on traditional advertising revenue. Each of these outlets vies for our attention by offering us not just an additional source of information, but a "better" source of information. There is not necessarily anything wrong with a broader variety of voices. Media consolidation (i.e., fewer unique voices) has long been a primary concern among scholars and media watchdog groups.[63] But this greater heterogeneity does provide an outlet, or perhaps an alibi, for those seeking to dismiss information that conflicts with their own opinions—a move implicitly or explicitly encouraged in an adversarial environment haunted by the specter of "fake news." The marketplace logic that governs the circulation of ideas and information within a democracy dictates that more ideas, more voices, are ultimately better for the individuals who make political choices based on this data. But the market logic that drives news outlets to compete for viewers through product differentiation based on the critique and dismissal of one's competitors seems to produce the opposite effect. As Yochai Benkler and his collaborators demonstrated, followers on both sides of the political spectrum in the 2016 US election were consuming a diet of media that came from across the political spectrum, but were consuming that information differently. Framed, reframed, and shared to one's network on Twitter and Facebook, these stories simply amplified or reinforced the opinions and input that one already possessed.[64] A broader diversity of media sources like the one provided by the midstream outlets, built on the premise that something is lacking in the news more generally, seems to produce dissensus and suspicion rather than consensus and trust in the media.

This broader diversity of media outlets also makes any level of brand recognition more difficult. Consider, for example, beyond the relatively larger midstream choices discussed so far, the broad array of choices and sources that confront users of digital news aggregators like Instapaper, Flipboard, Google News, and Apple News. These apps enable users to choose from and curate a broad diversity of sources, potentially enabling them to explore a variety of voices from across the increasingly politicized news spectrum.[65] Several studies have further demonstrated that these apps are also beneficial for news providers, bringing in additional readers who might not have been part of the existing audience.[66] While the experience of browsing such aggregators seems intended to remediate the experience of browsing titles at a newsstand in the analog era, the digital form dissolves any of the material indicators that might be used to judge the quality of a source's content. When getting something into print or even into a professionally produced website required vast resources, the mere presence of a news source spoke to some level of investment and commitment. This is no longer the case, and rather than a few trusted sources that one has tested over and over through years of consideration, we instead have an endless sea of choices and newcomers, all of which appear to be equal, at least on the surface. This makes for a deeper level of confusion about what's real or fake, trustworthy or unreliable. The diversity of sources provides the perfect camouflage for one site to slip in among the others, imitating the trappings and aesthetics of more-established sources in the way that *The Onion* or the Macedonian news sites did.

This sort of camouflage through imitation in the context of information density is what enabled the alleged disinformation campaigns that appeared on social media sites including Facebook, Twitter, and Instagram. In the midst of investigating potential Russian interference in the US election, the social media platforms provided a congressional committee with examples of the various propaganda.[67] Of what were purported to be thousands of advertisements purchased by these groups and paid for in rubles, fourteen examples were provided as a representative sample. As a whole, the aesthetics of these ads are fairly consistent, working to seed and spread the sort of viral, meme-driven media that already readily circulate in political channels across the web. Most seem to provide an alibi for their informal, unpolished look by pretending to be the work of amateurs and small grassroots organizations. To borrow Paul Arthur's term, this sloppy aesthetic is part of the rhetoric of authenticity, much in the way that shaky handheld camerawork connotes a type of amateur, documentary production.[68] Ideologically, they are all over the map, espousing contradictory and in some cases incomprehensible political positions, but all benefitting from the sort of oppositional, adversarial environment that predominates in this same mediascape. As their metadata indicate, none of them were seen by that many people, but most were shared forward multiple times beyond their original placement, making any assessment of their audience, and hence their impact, a near impossibility.

These advertisements thus offer a sort of stand-in or emblem of the larger environment of misinformation and disinformation, paranoia and suspicion that pervades any understanding of the US election in 2016. Their material plasticity, somewhere between advertisement, propaganda, and journalism, circulates in a space of unknowability: seen by an unidentifiable audience and produced and paid for by the same. Hidden within a much larger environment of media heterogeneity that feeds on critique and suspicion, they offer a marked contrast with the efforts at mass persuasion and propaganda at work in prior historical moments, representing instead the diffusion of media within a networked space. Purchased for small amounts of money, they seem to have slipped in unnoticed among everyone else on the back of a platform whose only logic is the exchange and monetization of content created by others. But like the forms of propaganda considered earlier, this is a feature of social networks like Facebook, not a bug.

Ironically, the emergence of all of these imitators, both real and fake, satirical and earnest, seems to have produced a positive outcome for the mainstream, traditional targets of their attack. Since the 2016 election and Donald Trump's one-man war on the media, many of these older news outlets have been asking for, and in many cases receiving, direct support from their readers in the form of sub-scriptions, donations, and other revenue streams that might replace the declining advertising revenue from their print and broadcast properties.[69] Almost in the way that the media critic Jean Baudrillard once claimed that Disneyland, with its simu-lations of Main Street USA and other fantastical, nostalgic spaces, provided an alibi or sense of reassurance that the real world outside the park still existed, these fake news websites and the paranoia and suspicion that they have engendered seem to have given traditional media a new aura of authenticity and importance.[70]

CONCLUSION

While the 2016 election and the surprise victory of Donald Trump may have been a shocking event that "shook the foundations of American politics," two of its most notable features—the cloud of suspicion and conspiratorial theorizing under which it unfolded and the reemergence of "fake news"—were the products of long-standing trends in American culture and its media.[71] The political backdrop—and, I would argue, accelerating cause—of these larger shifts was of course the war on terror. As the US government's response to 9/11 began to accentuate the political polarization and partisan infighting of the late 1990s, it also generated policies and practices seemingly designed to bear out, or perhaps simply validate, the darkest fears of the conspiracy community. These included things like CIA black sites, extraordinary rendition, John Poindexter's Total Information Awareness program, and at least rumors of the widespread surveillance practices that were eventually revealed by Edward Snowden. Snowden of course provided only one in a long line of explosive leaks that included those of WikiLeaks, of the Abu Ghraib images, and several other such leaks that quickly took on a role as fodder for conspiracy media in an ever-widening landscape of attempts to contain and explain the causal logic of these events. This culture of suspicion, and the perceived failure of traditional news outlets to adequately account for and address these events, in turn fueled the rise of the midstream news outlets discussed here.

In the early phases of the war on terror, political activists and independent doc-umentary filmmakers responded by utilizing newly available and rapidly evolving digital tools and channels of distribution as they sought to expose and condemn what many deemed to be an ill-conceived, unjust response to the terrorist at-tacks on September 11, 2001. The election of a progressive president on a netroots wave of enthusiasm touting a platform of "hope and change" brought with it the mainstream institutionalization of some of these technologies. These promises were further challenged by political activists and journalists seeking total trans-parency. The election of Donald Trump on a surge of nationalist populism not only redefined the political right in the United States but also demonstrated the extent to which the politics of opposition and the ethos of technological disruption had become part of the mainstream political discourse. By 2016, many of the tools utilized in these early official and oppositional responses to the war on terror (social networking, user-generated content, virtual simulation, data visualization) had become mainstream platforms in themselves. Just as sound technology or observational footage had evolved from cutting-edge, novel approaches into standard elements in the documentary lexicon, these new technologies became established frameworks for representing the world and distributing information. Where 2016 might have betrayed the seemingly democratizing, progressive potential that (some believed) was endemic to distributed digital technology, it also demonstrated that these newly evolved representational technologies were media like any other—media capable of both transparency and obfuscation, individual expression and mass manipulation, documentary alongside fiction. They had become a space, in short, where truth lies.

NOTES

1. SEEING IN THE DARK

The first of the quotations in the epigraph was uttered on the September 16, 2001, broadcast of the NBC show *Meet the Press,* in an interview with Tim Russert; the entire interview can be found in *Reining in the Imperial Presidency: Lessons and Recommendations Relating to the Presidency of George W. Bush,* a report prepared at the direction of Representative John Conyers Jr., Chairman of the US House of Representatives Committee on the Judiciary (Skyhorse Publishing, 2009). The second and third quotations are from, respectively, "Text of Obama's Speech in Afghanistan," *New York Times,* May 1, 2012, www.nytimes.com/2012/05/02/world/asia/text-obamas-speech-in-afghanistan.html; and "Remarks by President Trump at the Veterans of Foreign Wars of the United States National Convention | Kansas City, MO," accessed September 3, 2018, www.whitehouse.gov/briefings-statements/remarks-president-trump-veterans-foreign-wars-united-states-national-convention-kansas-city-mo/.

1. For a discussion of coverage of the air base and the alleged abuses that took place there, see Scott Horton, "Inside the Salt Pit," *Browsings: The Harper's Blog,* March 29, 2010, accessed April 18, 2013, http://harpers.org/blog/2010/03/inside-the-salt-pit/.

2. The original phrase appeared as a slogan in the film *Le vent d'est, 1970* (Groupe Dziga Vertov [Jean-Luc Godard, Jean-Pierre Gorin, Gérard Martin]). For a discussion, see Marcia Landy, "Just an Image: Godard, Cinema, and Philosophy," *Critical Quarterly* 43, no. 3 (2001): 9–31, https://doi.org/10.1111/1467-8705.00370.

3. Friedrich A. Kittler and Anthony Enns, *Optical Media: Berlin Lectures 1999* (Polity, 2010); Paul Virilio, *War and Cinema: The Logistics of Perception* (Verso, 1989), 71; Lisa Parks, *Cultures in Orbit: Satellites and the Televisual* (Duke University Press, 2005); Roger Stahl, *Militainment, Inc.: War, Media, and Popular Culture* (Routledge, 2009).

4. P. W. Singer, *Wired for War: The Robotics Revolution and Conflict in the 21st Century,* reprint (Penguin Books, 2009); Timothy Lenoir and Luke Caldwell, *The Military-Entertainment Complex* (Harvard University Press, 2017).

5. For a longer history of these rhetorical categories and the ends to which they are put within the frame of war, see Eric N. Olund, "Cosmopolitanism's Collateral Damage: The State-Organized Racial Violence of World War I and the War on Terror," in *Violent Geographies: Fear, Terror, and Political Violence,* ed. Derek Gregory and Allan Pred (Routledge, 2007). For a similar approach to the spatial redefinition of boundaries, see Amy Kaplan, "Where Is Guantanamo?," *American Quarterly* 57, no. 3 (October 10, 2005): 831–58, https://doi.org/10.1353/aq.2005.0048.

6. Michael M. Grynbaum, "Trump Calls the News Media the 'Enemy of the American People,'" *New York Times,* December 22, 2017, www.nytimes.com/2017/02/17/business/trump-calls-the-news-media-the-enemy-of-the-people.html.

7. The role of the "hacker" in digital culture offers an interesting figure for the consideration of activism and oppositional politics as a motor for technological change. For a discussion of both, see Alexander Galloway and Eugene Thacker, *The Exploit: A Theory of Networks* (University of Minnesota Press, 2007); and Tim Jordan, *Hacking: Digital Media and Technological Determinism* (Polity, 2008). For a discussion of the nuance involved in technological determinism in relation to Marx, see Donald A. MacKenzie, *Knowing Machines: Essays on Technical Change* (MIT Press, 1998), chap. 2. As Mackenzie demonstrates, the case isn't always as obvious as it may seem.

8. "Charting Documentary's Futures: An Interview with William Uricchio, Part Three," *MIT Open Documentary Lab* (blog), February 4, 2016, http://opendoclab.mit.edu/charting-documentarys-futures-an-interview-with-william-uricchio-part-three/.

9. By "worldview," I am referring not just to the more familiar usage that describes one's outlook or individual cosmological interpretation of events, although this is undoubtedly part of it. My conception of worldview encompasses not just the personal, social, and political ideology that we use to organize events and objects in the world but also the technological and media-driven means by which we come to know the world in the first place. In this sense I am attempting to marry the literal "frame" of the computer, television, tablet, and so on to the more abstract, psychological "framing effect" that any particular representation imposes. Documentary aesthetics in this sense not only instantiate a particular worldview but also shape and express the worldview of the audiences who see and agree with any particular film.

10. Michael Hardt and Antonio Negri, *Empire* (Harvard University Press, 2000); Wendy Brown, *Walled States, Waning Sovereignty* (Zone Books, 2010); Wendy Brown, *Undoing the Demos: Neoliberalism's Stealth Revolution,* reprint (Zone Books, 2017); McKenzie Wark, *A Hacker Manifesto* (Harvard University Press, 2004).

11. Jacques Rancière, *The Politics of Aesthetics,* trans. Gabriel Rockhill (Bloomsbury Academic, 2006), 12.

12. A brief note here on terminology: the two media listed here are enormously capacious and lend themselves to widely divergent interpretations in different contexts. "The moving image," of course, can refer to both film and television, two separate forms of media that have unique histories and disciplines, and yet the two overlap in that they share similar properties, producers, and content. My point here is not that all forms of moving images are

identical (nor are all Internets or digital technologies), but simply that, taken together, these general groups point to larger historical trends.

13. Douglas Gomery, *The Coming of Sound* (Routledge, 2005), 14–20; Donald Crafton, *The Talkies: American Cinema's Transition to Sound, 1926–1931,* vol. 4 (University of California Press, 1999); Lynn Spigel, *Make Room for TV: Television and the Family Ideal in Postwar America* (University of Chicago Press, 1992); William Boddy, "The Studios Move into Prime Time: Hollywood and the Television Industry in the 1950s," *Cinema Journal* 24, no. 4 (1985): 23–37, https://doi.org/10.2307/1224894.

14. To see this sort of transitional integration between technologies at work across a broad range of media, consider Jay David Bolter and Richard Grusin, *Remediation: Understanding New Media* (MIT Press, 2000). Marshall McLuhan, of course, offers the prototypical model for this in *Understanding Media: The Extensions of Man* (McGraw-Hill, 1965), which Bolter and Grusin reference in their title.

15. In this sense my study draws on the model of media emergence Bolter and Grusin outline in their influential work *Remediation.* They describe the process as a double action in which any new form of media both draws on and influences prior forms of media. Thus, a new technology like the web will draw from existing media like newspapers (in both form and content) in order to gain legitimacy, while at the same time these older forms will adopt characteristics of new forms in order to remain viable in this shifting media landscape. In my formulation, new forms of networked political action will thus draw on certain tendencies and characteristics of documentary film even as documentary itself evolves to take advantage of newly available digital technologies. Given this, my aim is to chart a spectrum of different work, from mainstream political documentaries on one end to those that bear little actual resemblance to documentary on the other.

16. This independent tradition, as Patricia Zimmerman's powerful account *States of Emergency: Documentaries, Wars, Democracy* (University of Minnesota Press, 2000) demonstrates, was one of the targets that emerged in the decade or so between the end of the Cold War and the start of the war on terror. Zimmermann meticulously outlines the connections between public media, public spaces, and the independent voices and agencies that act as a check on power. Zimmermann's own work through festivals (primarily the Finger Lakes Environmental Film Festival), the Flaherty Seminar, the Visible Evidence community, and many other venues is another example of the deep connections between independent documentary and political activism.

17. This has been extensively discussed by both Bill Nichols and Carl Plantinga, among others. See Bill Nichols, "Rhetoric and What Exceeds It," chap. 5 in *Representing Reality: Issues and Concepts in Documentary* (Indiana University Press, 1991); and Carl R. Plantinga, *Rhetoric and Representation in Nonfiction Film* (Cambridge University Press, 1997).

18. Not all of the blame for this extreme polarization can be laid at Bush's feet, however. The protests that emerged during the World Trade Organization meetings in Seattle, the Clinton impeachment over the Monica Lewinsky scandal, and Newt Gingrich's "Contract with America" in 1995 might all be cited as further starting points for a political trend that has hardly ceased since Bush left office. Wherever we place the starting point, however, it should still be noted that events during Bush's tenure pushed this general political enmity to a fevered pitch.

19. As Jane Mayer notes, the debate that I refer to here wasn't so much an internal debate about the ethics of torture within the Bush administration, but rather an attempt to stretch semantic and legal boundaries in order to accommodate the inhumane and illegal techniques already often in use by the CIA and other groups in charge of interrogating suspects. In any case, most observers ended up agreeing with Mark Danner's point of view that regardless of how you categorize these illegal and inhumane acts, they are also ineffective, rendering any payoff irrelevant. See Jane Mayer, *The Dark Side: The Inside Story of How The War on Terror Turned into a War on American Ideals* (Knopf Doubleday, 2009), 148–57; and Mark Danner, *Torture and Truth: America, Abu Ghraib, and the War on Terror* (New York Review of Books, 2004), 10–26.

20. José van Dijck and her later collaborators have brilliantly delineated the evolution of these impulses into their eventual commercialization on some of the largest platforms on the web today (Facebook, Twitter, Google, Uber). See José van Dijck, *The Culture of Connectivity: A Critical History of Social Media* (Oxford University Press, 2013); and José van Dijck, Thomas Poell, and Martijn de Waal, *The Platform Society* (Oxford University Press, 2018). Clay Shirky provides a number of useful case studies demonstrating the productive power of these tools. See Clay Shirky, *Here Comes Everybody: The Power of Organizing without Organizations,* reprint (Penguin, 2009).

21. It would be more accurate to say that Snowden's revelations accelerated these debates, given that the NSA's warrantless wiretapping program was already an object of clear scrutiny by many. For a full account of both, see Glenn Greenwald, *No Place to Hide: Edward Snowden, the NSA, and the U.S. Surveillance State* (Henry Holt and Company, 2014).

22. Any number of examples might be included here, from the canonical account Erik Barnouw includes in his *Documentary: A History of the Non-Fiction Film* (Oxford University Press, 1974) to Brian Winston's discussion in "The Documentary Film as Scientific Inscription," in Michael Renov, *Theorizing Documentary* (Routledge, 1993) to the more recent book-length discussion in Dave Saunders, *Direct Cinema* (Wallflower Press, 2007).

23. The same scenario plays out again with the transition from film to video and the mainstream emergence of cable television and the Public Broadcasting System (PBS) in the 1970s and 1980s. The decreased production cost of shooting on video enabled an explosion of new documentary forms, from the autobiographical work of people like Sadie Benning, Abraham Ravett, and Alan Berliner to the political documentaries of other filmmakers like Jill Godmilow, Rea Tajiri, Marlon Fuentes, and Marlon Riggs. Like the major television networks that first exhibited the work of many Direct Cinema filmmakers, PBS and public access both proved to be important distributors and exhibition sites for documentary work produced on video. As with Direct Cinema, changes in technology accompanied changes in the form, subject, and audience of documentary film. To borrow the taxonomy that Bill Nichols puts forward, we might simply state that just as 16mm and network television in the 1950s and 1960s gave rise to the "observational" and "participatory" modes of documentary, video and public television gave rise to the "performative" mode. See Nichols, *Representing Reality*.

24. Kevin McDonald and Daniel Smith-Rowsey, *The Netflix Effect: Technology and Entertainment in the 21st Century* (Bloomsbury Academic, 2016); "The Shocking Truth," *The Economist,* August 27, 2013, www.economist.com/prospero/2013/08/27/the-shocking-truth.

25. Quoted in Richard Roud, *Jean-Luc Godard* (Thames and Hudson/British Film Institute, 1970), 163. Significantly, his remark is directed at Richard Leacock's assertion that Direct Cinema's technological enhancements enabled it to foster subject-driven observation.

26. Indeed, cheaper access to production materials, as Jennifer Malkowski notes in her brilliant exploration of death (documentary's most taboo of topics), allows many more stories to be told. Quoting Kate Davis, the director of *Southern Comfort,* she notes that this, and a great many other stories, might have failed to be told if they had depended upon the interest of traditional funding outlets or large crews in order to get started. See Jennifer Malkowski, *Dying in Full Detail: Mortality and Digital Documentary* (Duke University Press, 2017), 88.

27. The responses to the specter of technological determinism run the gamut from those from Friedrich Kittler, who openly embrace it as a motor of history, to those who posit social and cultural forces as the motor of technological development, as Brian Winston does. A typically neutral position is offered in the work of Carolyn Marvin. See Friedrich A. Kittler, *Gramophone, Film, Typewriter* (Stanford University Press, 1999); Brian Winston, *Media Technology and Society* (Routledge, 1998); and Carolyn Marvin, *When Old Technologies Were New* (Oxford University Press, 1990).

28. The term "performative" comes from the extension of linguistic performatives (J. L. Austin and Judith Butler) to nonfiction film expression. See Bill Nichols, *Blurred Boundaries* (Indiana University Press, 1994); Stella Bruzzi, *New Documentary* (Routledge, 2000), chap. 5, "Performing Documentary"; and, for the term "on/scene," Linda Williams, *Porn Studies* (Duke University Press, 2004).

29. Jerry Kuehl, "Truth Claims," in *New Challenges for Documentary,* ed. Alan Rosenthal (University of California Press, 1988).

30. Jonathan Kahana, *Intelligence Work: The Politics of American Documentary* (Columbia University Press, 2008).

31. See Michael Renov, "Early Newsreel: The Construction of a Political Imaginary for the New Left," chap 1. in *The Subject of Documentary* (University of Minnesota Press, 2004).

32. Hayden White, *The Content of the Form: Narrative Discourse and Historical Representation* (Johns Hopkins University Press, 1990).

33. See, e.g., Trinh T. Minh-ha's films *Reassemblage* (1983) and *Surname Viet Given Name Nam* (1989) and either of her books *Framer Framed: Film Scripts and Interviews* (Routledge, 1992) and *Woman, Native, Other: Writing Postcoloniality and Feminism* (Indiana University Press, 1989).

34. The move I am describing here can be seen in work by Paper Tiger TV, Chris Marker, Harun Farocki, Lynne Sachs, Péter Forgács, Leandro Katz, Jeffrey Skoller, and others. All of these artists demonstrate in one way or another the connections between an expressive aesthetic form and a committed engagement with politics.

35. Kahana's *Intelligence Work,* which reiterates, incidentally, the same three periods outlined above, deals in particular with films after the emergence of postmodern skepticism but before 9/11. His reading of *Journeys with George,* Alexandra Pelosi's 2000 documentary about the first Bush election, characterizes ideally the situation of political ennui indicative of the pre-9/11 documentary.

36. One, admittedly limited, indication of this increased engagement is the twelve-year shift from 1996's record low voter turnout to 2008's nearly record-breaking turnout.

For a discussion of this, with statistics, see "Election Drew Highest Turnout in 40 Years," *The Caucus* (blog), December 15, 2008, https://thecaucus.blogs.nytimes.com/2008/12/15/election-drew-highest-turnout-in-40-years/.

37. Charles Musser, "Film Truth in the Age of George W. Bush," *Framework: The Journal of Cinema and Media* 48, no. 2 (2007).

38. The importance of the Internet when combined with the DVD as a distribution channel cannot be overestimated. Not only does this cement documentary's claim to providing an "independent" point of view outside of the mainstream media, but it also allows documentary films that would have lacked the resources to mount a theatrical release or television commitment from a major network (all of whom, including PBS, were increasingly less willing to give time to "political" films for fear of being branded organs of liberal/conservative media) to eventually find a niche market in the "long tail" (the term given to the large number of products and films that find small audiences instead of the small number that find large audiences) online.

39. A note here on terminology. Although the phrase "database aesthetics" was formalized with the recent publication of Victoria Vesna's edited volume *Database Aesthetics: Art in the Age of Information Overflow* (University of Minnesota Press, 2007), a number of uses had sprung up prior to this—most notably for me here, in Lev Manovich, "Database as Symbolic Form," *Convergence: The International Journal of Research Into New Media Technologies* 5, no. 2 (June 1999): 80–99. An extended discussion of the term forms part of this chapter, particularly since my reading here stretches its usage somewhat by arguing that a film text (Manovich's primary example of narrative aesthetics and, hence, the binary opposite of the database) is used by Morris to overtly explore the database aesthetic at work in a collection of digital images.

40. Johan Huizinga, *Homo Ludens: A Study of the Play-Element in Culture* (Beacon Press, 1955).

41. A few examples: DNA provides the "code" of life; we can track our health and well-being through various personal metrics and record all of our relationships on social networks that in turn provide a "social graph" of our most intimate connections; the health of the state is measured and administered through various statistical indicators whose models attempt to optimize results through various "tweaks" to the data.

2. "WE SEE WHAT WE WANT TO BELIEVE"

The epigraph comes from a review by Jorge Luis Borges of Orson Welles's *Citizen Kane*. See www.openculture.com/2014/08/jorge-luis-borges-film-critic-reviews-citizen-kane.html.

1. See www.errolmorris.com.

2. As in all of his films, Morris is still interested in the way that given individuals arrive at the startlingly idiosyncratic and erroneous conclusions that they do, but in these two recent films, his focus is more directly on the media-laden road that they took to get there. See, for example, his 2000 interview in the journal *Cineaste* about his film *Mr. Death,* where he states, "Ultimately, it is a movie about denial. Denial about the obvious, denial of self, denial of death, denial of the Holocaust. But at its center, it is a failure to see the world, to see reality. Living in a cocoon of one's own devising. Fabricating

a universe that one occupies, that may in fact be divorced from the real world." Roy Grundmann and Cynthia Rockwell, "Truth Is Not Subjective: An Interview with Errol Morris," *Cineaste* 25, no. 3 (January 2000): 6.

3. The transmediation of analog material to digital archives on the Internet has been a topic of concern both for those actually doing the work (librarians, archivists, and others) and for New Media scholars concerned with the theoretical frameworks that ground such projects. Wendy Chun, for example, has pointed out that memory and storage are often erroneously conflated in progressivist accounts of the Internet's ability to save analog media while providing complete, instant, permanent access to the world's information. Marija Dalbello, writing as a librarian and archivist, worries on the other hand that digital archives as currently instantiated don't present a complete enough historical record, focusing as they do on the popularly accessible collection over the complete and rigorous approach. See Wendy Hui Kyong Chun, "The Enduring Ephemeral, or the Future Is a Memory," *Critical Inquiry* 35, no. 1 (Autumn 2008): 148–71; and Marija Dalbello, "Institutional Shaping of Cultural Memory: Digital Library as Environment for Textual Transmission," *Library Quarterly* 74, no. 3 (July 2004): 265–98.

4. Robert S. McNamara and James G. Blight, *Wilson's Ghost: Reducing the Risk of Conflict, Killing, and Catastrophe in the 21st Century* (PublicAffairs, 2003). For a discussion of the collaboration between McNamara, the Wilson Institute, and the authors of the film's companion book, see James G. Blight and janet M. Lang, *The Fog of War: Lessons from the Life of Robert S. McNamara* (Rowman & Littlefield, 2005), 9.

5. The Interrotron is Morris's signature, self-invented camera setup for conducting interviews. It essentially consists of two modified teleprompters placed in separate rooms, which each project the feed from a camera placed behind the other. This allows the subject to look Morris, and consequently the camera, in the eye as the interview is conducted. On-screen in the final film, the setup produces an unsettling degree of eye contact between subject and viewer.

6. On a side note, "Lesson 1: Empathize with Your Enemy," which draws on McNamara's experiences with both Fidel Castro and Nikita Khrushchev, also seems to double as an enjoinder to the audience, many of whom might vilify McNamara for his role in Vietnam, to check any prejudices against McNamara at least long enough to hear him out.

7. Paul N. Edwards, *The Closed World: Computers and the Politics of Discourse in Cold War America* (MIT Press, 1996), http://hdl.handle.net/2027/heb.01135.

8. Morris, *The Fog of War* (2003), 2:55–3:10.

9. See James G. Blight and janet M. Lang, "Lesson 2: Rationality Will Not Save Us," chap. 2 in Blight and Lang, *The Fog of War;* and McNamara and Blight, *Wilson's Ghost,* 237–44.

10. Blight and Lang, *The Fog of War,* 59.

11. The show originally aired on September 25, 1963, on CBS.

12. Morris, *The Fog of War* (2003), 4:57–5:41.

13. Ibid., 28:17.

14. As Vinzenz Hediger points out in his discussion of Hollerith in early German industrial films, "Both technologically and in institutional terms, the Hollerith data processor is a predecessor of the modern-day computer. The Hollerith tabulating machine uses punched cards to tabulate statistics from data" (133). See Vinzenz Hediger and Patrick Vonderau, *Films That Work: Industrial Film and the Productivity of Media* (Amsterdam University Press, 2009).

15. The controversy to which I refer was set off by the publication of Edwin Black's book detailing the relationship between IBM and its German subsidiary Dehomag. Black, a science fiction writer, has steadfastly defended the book's claims while denouncing its detractors as corporate thugs for IBM. The book was both widely lauded and condemned, achieving best-seller status while earning Black awards for his work as well as condemnation as a crank. See Edwin Black, *IBM and the Holocaust: The Strategic Alliance between Nazi Germany and America's Most Powerful Corporation* (Dialog Press, 2008). For refutation of the book and an outline of its faults, see Michael Allen, "Review: Stranger than Science Fiction: Edwin Black, IBM, and the Holocaust," *Technology and Culture* 43, no. 1 (January 2002): 150–54.

16. Norbert Wiener's role in creating military targeting systems under the direction of Vannevar Bush led him to conceptualize future human–machine relationships in the nascent field of cybernetics. See his "Men, Machines, and the World About," in *The New Media Reader*, ed. Noah Wardrip-Fruin and Nick Montfort (MIT Press, 2003).

17. See Vannevar Bush, "As We May Think," in *The New Media Reader*, ed. Noah Wardrip-Fruin and Nick Montfort (MIT Press, 2003).

18. Wiener's ethical stance on government funding is still commemorated by the annual Wiener Award, which the Computer Professionals for Social Responsibility group hands out. See Wiener , "Men, Machines," 65.

19. See http://www.mtv.com/news/2759357/an-interview-with-errol-morris-director-of-standard-operating-procedure/ and Taylor Segrest, "2009 Career Achievement Award: The Cinematic Investigations of Errol Morris," www.documentary.org/feature/2009-career-achievement-award-cinematic-investigations-errol-morris.

20. See Terry Gross, "Interview with Errol Morris," *Fresh Air with Terry Gross*, January 5, 2004, http://39ea54ff11b298f9bcaa-1b99eba380497722926169d6da8b098e.r2.cf5.rackcdn.com/2004/FA20040105.mp3.

21. Morris, *The Fog of War* (2003), 1:06:10–1:10:53.

22. The "domino theory," as it came to be known, held that if one country fell to communist control, the other countries in Southeast Asia would quickly follow. As Gordon Goldstein puts it: "By 1964 the domino theory had the force of doctrine, becoming a de facto feature of the political debate over Vietnam, the teetering domino that could ostensibly unleash communism across Southeast Asia." Gordon M. Goldstein, *Lessons in Disaster: McGeorge Bundy and the Path to War in Vietnam* (Macmillan, 2008), 139.

23. Blight and Lang, *The Fog of War*, 90.

24. This of course is hardly the only account of these events, or of their interpretation. Eric Alterman's account of the incident describes Johnson as being "deliberately deceptive" about the event in order to pass the Gulf of Tonkin Resolution, which gave him the unconditional authority to escalate the war. See his chapter titled "Lyndon B. Johnson and the Gulf of Tonkin Incidents," in Eric Alterman, *When Presidents Lie: A History of Official Deception and Its Consequences* (Penguin, 2005).

25. Indeed, David Mosen's analysis of *Report* in *Film Quarterly* sounds eerily reminiscent of *The Fog of War*. He states: "In Conner's eyes society thrives on violence, destruction, and death no matter how hard we try to hide it with immaculately clean offices, the worship of modern science, or the creation of instant martyrs. From the bullfight arena to the nuclear arena we clamor for the spectacle of destruction." David Mosen, "Review: 'Report' Bruce

Conner," *Film Quarterly* 19, no. 3 (Spring 1966): 55. For an extended analysis of the film, which expands upon the connection between the film and its media roots (what he calls the "complicity of the moving image media in the rise and fall of John F. Kennedy" [250]), see Bruce Jenkins, "Bruce Conner's *Report*: Contesting Camelot," in *Masterpieces of Modernist Cinema*, ed. Ted Perry (Indiana University Press, 2006), 236–51.

26. In the interview with arts blogger Greg Allen, Morris actually claims that "dozens" of such stills were created for the film, but that this is the only one that survived into the final cut. Greg Allen, *Greg.Org: The Making of* (blog), "Learning at Errol Morris's Knee," accessed September 19, 2010, http://greg.org/archive/2004/02/20/learning_at_errol_morriss_knee.html.

27. Consider, for instance, the recent evolution and ongoing debates in the discipline of history itself. While the field is hardly unified into a single body of practitioners espousing a single set of concerns, several recent texts might be pointed to as evidence of a larger disciplinary evolution, including: (1) the rise and fall of both social and cultural history as chronicled in chapter 2 of William Sewell's *The Logics of History* (University of Chicago Press, 2009) and more fully elaborated in Peter Burke's *New Perspectives on Historical Writing* (Pennsylvania State University, 2011); (2) the effects of poststructuralism and the larger linguistic turn as explored in Frank Ankersmit and Hans Kellner's edited volume *A New Philosophy of History* (University of Chicago Press, 1995) and Derek Attridge's *Post-Structuralism and the Question of History* (Cambridge University Press, 1999); and (3), more generally, the ongoing scandal incited by the publication of Hayden White's *Metahistory* (Johns Hopkins University Press, 1987), or the overall influence of the French Annales school.

28. Grundmann and Rockwell, "Truth Is Not Subjective." 4.

29. Ibid., 7.

30. This lack of transparency regarding the past, and the ability to misinterpret evidence, explains in part Morris's continued fascination with those individuals who delude themselves for whatever reason about the information they are presented with, be it the Dallas Police Department (*The Thin Blue Line*), Fred Leuchter (*Mr. Death*), or Robert McNamara.

31. Grundmann and Rockwell, "Truth Is Not Subjective," 7.

32. Jacques Derrida, *Archive Fever: A Freudian Impression* (University of Chicago Press, 1995).

33. Here we could point to such high-profile endeavors as the Google Books's scanning project, which seeks to create a digital copy of every book ever printed, on down to the activities of individual libraries and archives that work on specific collections of photographs and historical documents, to subject-focused archives like the Trans-Atlantic Slave Trade Database, which consolidates and digitizes historical records from a number of historical sources relating to the transatlantic slave trade. Other notable examples include the Archive.org Prelinger Archives, which digitizes and distributes found films, and journal and newspaper databases such as the popular JSTOR project or the *New York Times* archive, which makes every article published in the *New York Times's* 150-year history searchable online.

34. Chun, "The Enduring Ephemeral." 148-171.

35. Vannevar Bush, "As We May Think," *The Atlantic*, July 1945, www.theatlantic.com/magazine/archive/1945/07/as-we-may-think/303881/; John von Neumann, "First Draft of a Report on the EDVAC," *IEEE Annals of the History of Computing* 15, no. 4 (1993): 27–75, https://doi.org/10.1109/85.238389.

36. Ibid., 20.

37. Stephen Bann, "History as Competence and Performance: Notes on the Ironic Museum," in Ankersmit and Kellner, *A New Philosophy of History*, 210.

38. Ibid., 200–210.

39. Pat Joseph, "The World According to Errol Morris," *California Magazine*, Spring 2010, 29.

40. Linda Williams, "Cluster Fuck: The Forcible Frame in Errol Morris's Standard Operating Procedure," *Camera Obscura* 25, no. 1 (2010): 33.

41. See, for example, the reviews by Manohla Dargis in the *New York Times* and Jay Hoberman in the *Village Voice*. Ironically, Morris's budget for *Standard Operating Procedure* was $5 million, roughly the amount that *The Fog of War* had grossed, but went on to make only a scant $350,000 in worldwide release. See Manohla Dargis, "We, the People Behind the Abuse," *New York Times*, April 25, 2008, www.nytimes.com/2008/04/25/movies/25stan.html; and Jay Hoberman, "Errol Morris Lets Torturers Off Easy," *Village Voice*, April 22, 2008, www.villagevoice.com/2008-04-22/film/get-out-of-jail-free/.

42. In spite of contentious discussions of the film at the 2009 Visible Evidence conference and the Society for Cinema and Media Studies conference in 2010 (both in Los Angeles) between parties who defended and attacked the film, it seems that only its proponents have taken the additional step of publishing their thoughts on the film. See, for example, Williams, "Cluster Fuck"; Charles Musser, "Political Documentary, YouTube and the 2008 US Presidential Election: Focus on Robert Greenwald and David N. Bossie," *Studies in Documentary Film* 3, no. 3 (2009): 199–218; Hilary Neroni, "The Nonsensical Smile of the Torturer: Documentary Form and the Logic of Enjoyment," *Studies in Documentary Film* 3, no. 3 (2009): 245–57; Julia Lesage, "Torture Documentaries," *Jumpcut: A Review of Contemporary Media* 51 (Spring 2009), www.ejumpcut.org/archive/jc51.2009/TortureDocumentaries/; and Caetlin Benson-Allott, "Standard Operating Procedure: Mediating Torture," *Film Quarterly* 62, no. 4 (Summer 2009): 39–44.

43. Philip Gourevitch and Errol Morris, *Standard Operating Procedure: A War Story* (Picador, 2013).

44. Raymond Bonner, "The Dogs of War," *New York Times*, May 25, 2008, www.nytimes.com/2008/05/25/books/review/Bonner-t.html.

45. Consider, for example, the difference in the two companion volumes. The Blight and Lang companion text for *The Fog of War* offers something like a set of footnotes to the historical documents and other materials that the film references in passing. In essence, it substantiates the claims the film makes with an expanded set of materials as evidence. Gourevitch's narrative for *Standard Operating Procedure* is something closer to a complete story or expanded version of the events in the film. Whereas the ancillary media on errolmorris.com for *The Fog of War* are directly about the film, the ancillary media for *Standard Operating Procedure* extend the issues discussed in the film.

46. Errol Morris, "Liar, Liar, Pants on Fire," *Opinionator* (blog), July 10, 2007, http://opinionator.blogs.nytimes.com/2007/07/10/pictures-are-supposed-to-be-worth-a-thousand-words/.

47. Errol Morris, "Cartesian Blogging, Part One," *Opinionator* (blog), December 10, 2007, http://opinionator.blogs.nytimes.com/2007/12/10/primae-objectiones-et-responsio-auctoris-ad-primas-objectiones-part-one/.

48. Errol Morris, "The Anosognosic's Dilemma: Something's Wrong but You'll Never Know What It Is (Part 1)," *Opinionator* (blog), June 10, 2010, http://opinionator.blogs.nytimes.com/2010/06/20/the-anosognosics-dilemma-1/.

49. See Susan Sontag, *Regarding the Pain of Others* (Macmillan, 2004).

50. Ulrich Keller, *The Ultimate Spectacle: A Visual History of the Crimean War* (Gordon and Breach, 2001).

51. Irina Leimbacher, for example, claims that the reenactments are "a subject of fetishistic display or perverse ornamental possession for Morris." See Irina Leimbacher, "Response to Papers and Comments on *Standard Operating Procedure*," *Jumpcut: A Review of Contemporary Media* 52 (Summer 2010), www.ejumpcut.org/archive/jc52.2010/sopLimbacher/index.html.

52. As Linda Williams puts it, "[T]he humiliation, abuse and torture are so often enacted as if *for* the camera." Williams, "Cluster Fuck," 48.

53. Errol Morris, "Which Came First? (Part Two)," *Opinionator* (blog), October 4, 2007, http://opinionator.blogs.nytimes.com/2007/10/04/which-came-first-part-two/.

54. These three films obviously place *Standard Operating Procedure* in a constellation of films on torture and detainee abuse. Read alternatively as a film about the direction of the war in Iraq, it might be seen alongside others like *Iraq in Fragments* (James Longley, 2006), *Gunner Palace* (Petra Epperlein and Michael Tucker, 2004), *Uncovered: The War on Iraq* (Robert Greenwald, 2004), and *No End in Sight* (Charles Ferguson, 2007).

55. I place the term "like" in scare quotes simply because it is hard to claim that one likes a film about such a horrific topic. It should be taken therefore to stand in for something more akin to my earlier formulation—namely, that one does or does not find merit in the film's approach.

56. The controversy to which I am alluding here was nicely formalized in a panel presentation at the 2010 Society for Cinema and Media Studies conference in Los Angeles, which brought together scholars on both sides of the issue. Their points were later printed in issue 52 of the journal *Jump Cut*. On the side of the film were Jonathan Kahana and Linda Williams, with Nichols and Leimbacher both criticizing it. The discussion afterward largely mirrored the split on the panel. See "Conference Report: Reframing *Standard Operating Procedure*—Errol Morris and the Creative Treatment of Abu Ghraib," *Jump Cut* 52 (Summer 2010).

57. Bill Nichols, "Feelings of Revulsion and the Limits of Academic Discourse," *Jump Cut* 52 (Summer 2010), www.ejumpcut.org/archive/jc52.2010/sopNichols/index.html.

58. Lesage, "Torture Documentaries."

59. Williams, "Cluster Fuck."

60. Benson-Allott, "Standard Operating Procedure," 41.

61. The term is taken from Victoria Vesna's new collection of the same name. In it, Vesna collects discussions from several prominent New Media artists as well as theorists like Warren Sack and Lev Manovich, and curators of digital art like Christiane Paul. See Victoria Vesna, ed., *Database Aesthetics* (University of Minnesota Press, 2007).

62. Lev Manovich, *The Language of New Media* (MIT Press, 2001).

63. The term "Universal Media Machine" is Manovich's, one he uses to describe the rise to prominence of the computer as an essential tool in the creation, distribution, and consumption of various forms of media.

64. It should be noted that this does not mean that the database itself is unstructured, but simply that interacting with a database from an end-user perspective is open to many different structures and interpretations.

65. These sequences contain the film's most extensive use of computer-generated imagery.

66. W. J. T. Mitchell, *Cloning Terror: The War of Images, 9/11 to the Present*, (University of Chicago Press, 2011), 204.

67. Morris, "Liar, Liar."

68. *The Economist*, May 8, 2004.

69. Here I am thinking of *The Thin Blue Line*, which has political implications in its critique of the justice system, as well as *Mr. Death*, which deals with the politics around anti-Semitism and Holocaust denial. But where one explores these issues through an isolated event (the false accusation of Randall Adams in *The Thin Blue Line*) and the other through an individual figure (Fred Leuchter in *Mr. Death*), both *Standard Operating Procedure* and *The Fog of War* deal with issues (Abu Ghraib, Vietnam) that were a primary focus of national discussion over an extended period.

3. NETWORKED AUDIENCES

The epigraph is from Clay Shirky, *Here Comes Everybody: The Power of Organizing without Organizations*, reprint (Penguin, 2009).

1. Trending BBC, "'Pizzagate': The Fake Story That Shows How Conspiracy Theories Spread," *BBC News*, December 2, 2016, www.bbc.co.uk/news/blogs-trending-38156985; Cecilia Kang, "Fake News Onslaught Targets Pizzeria as Nest of Child-Trafficking," *New York Times*, December 22, 2017, www.nytimes.com/2016/11/21/technology/fact-check-this-pizzeria-is-not-a-child-trafficking-site.html; Adam Goldman, "The Comet Ping Pong Gunman Answers Our Reporter's Questions," *New York Times*, January 20, 2018, www.nytimes.com/2016/12/07/us/edgar-welch-comet-pizza-fake-news.html.

2. Thomas Waugh, ed., *"Show Us Life": Toward a History and Aesthetics of the Committed Documentary* (Scarecrow Press, 1984).

3. Theodore Hamm, *The New Blue Media: How Michael Moore, MoveOn.Org, Jon Stewart and Company Are Transforming Progressive Politics* (New Press, 2008).

4. MoveOn.Org, "MoveOn.Org: About," https://front.moveon.org/about/; "MoveOn.Org," Ballotpedia, accessed November 20, 2018, https://ballotpedia.org/MoveOn.org.

5. The eventual interconnection between moving images and social technology seems to provide further evidence of the "convergence culture" that Henry Jenkins and others describe. See Henry Jenkins, *Convergence Culture* (NYU Press, 2006). Instead, I'll argue that documentary's long connection to social change pushed it to prominence at a time when the utopian hopes it inspired were increasingly placed instead on newer digital forms of social organization.

6. Robert S. Boynton, "How to Make a Guerrilla Documentary," *New York Times Magazine*, July 11, 2004, www.nytimes.com/2004/07/11/magazine/how-to-make-a-guerrilla-documentary. html?pagewanted = 1.

7. Charles Musser, "Political Documentary, YouTube and the 2008 US Presidential Election: Focus on Robert Greenwald and David N. Bossie," *Studies in Documentary*

Film 3, no. 3 (December 2009): 199–218, https://doi.org/10.1386/sdf.3.3.199/1; Christian Christensen, "Political Documentary, Online Organization and Activist Synergies," *Studies in Documentary Film* 3, no. 2 (November 2009): 77–94, https://doi.org/10.1386/sdf.3.2.77/1. For Musser, Greenwald was ahead of the curve (and, significantly, ahead of conservative activists working toward the same end) on moving from the traditional theatrical release first to DVDs and then to YouTube as a means of allowing greater access to his films in critical periods before the US presidential elections in 2004, 2006, and 2008.

8. Christensen, "Political Documentary."

9. Victoria Carty, "New Information Communication Technologies and Grassroots Mobilization," *Information, Communication & Society* 13, no. 2 (2010): 155, https://doi.org/10.1080/13691180902915658; Victoria Carty and Jake Onyett, "Protest, Cyberactivism and New Social Movements: The Reemergence of the Peace Movement Post 9/11," *Social Movement Studies: Journal of Social, Cultural and Political Protest* 5, no. 3 (2006): 229, https://doi.org/10.1080/14742830600991586; Marc Eaton, "Manufacturing Community in an Online Activist Organization: The Rhetoric of MoveOn.org's E-Mails," *Information, Communication & Society* 13, no. 2 (2010): 174, https://doi.org/10.1080/13691180902890125; Paul Ferber, "Cyberdemocracy and Online Politics: A New Model of Interactivity," *Bulletin of Science, Technology & Society* 27, no. 5 (October 2007): 391–400; Noriko Hara and Zilia Estrada, "Analyzing the Mobilization of Grassroots Activities via the Internet: A Case Study," *Journal of Information Science* 31, no. 6 (December 2005): 503–14, https://doi.org/10.1177/0165551505057013.

10. See, for example, Brian Stelter, "Released on Web, a Film Stays Fresh," *New York Times,* March 23, 2009, http://www.nytimes.com/2009/03/23/movies/23gree.html.

11. Robert Greenwald, "Robert Greenwald's Documentaries," accessed January 27, 2011, www.robertgreenwald.org/docs.php.

12. Musser, "Political Documentary," 201.

13. Robert Greenwald, "*UNCOVERED: The War on Iraq:* The Director's Introduction," https://web.archive.org/web/20051126020837/http://www.truthuncovered.com/introduction.php.

14. Hamm, *New Blue Media.*

15. Bill Nichols, *Newsreel: Documentary Filmmaking on the American Left* Arno Press, 1980); Deedee Halleck and Nathalie Magnan, "Access for Others: Alter(Native) Media Practice," *Visual Anthropology Review* 9, no. 1 (March 1993): 154–63; Bill Nichols, "The American Photo League," *Screen* 13, no. 4 (1972): 108.

16. Greenwald, "UNCOVERED."

17. Boynton, "Guerrilla Documentary."

18. Stelter, "Released on Web."

19. Musser, "Political Documentary." As Musser points out, it was only with the appearance of YouTube in 2006 that a turnkey solution for online video streaming was made available to a wide audience of producers and consumers, making it a natural outlet for those on both sides of the media equation seeking a way around what Greenwald referred to as "the traditional gatekeepers."

20. Christensen, "Political Documentary."

21. Katie Hafner, "Mobilizing on Line for Gun Control," *New York Times,* May 20, 1999, www.nytimes.com/1999/05/20/technology/screen-grab-mobilizing-on-line-for-gun-control.html.

22. Robert Cwiklik, "MoveOn Targets Campaign 2000 Voters Web Fund-Raiser Tries to Revive Impeachment Angst," *Wall Street Journal*, February 2, 2000.

23. Alex Jordanov and Scott Stevenson, *MoveOn: The Movie*, DVD (Brave New Films, 2009).

24. Ian Stewart, "Anti-War Group Revives LBJ-Era 'Daisy' Ad Campaign," Associated Press, January 15, 2003.

25. Jordanov and Stevenson, *MoveOn: The Movie*.

26. "Bush in 30 Seconds," accessed February 2, 2011, http://www.bushin30seconds.org/.

27. Sandy Brown, "MoveOn Denied Super Bowl Slot," *AdWeek*, January 25, 2004, www.adweek.com/brand-marketing/moveon-denied-super-bowl-slot-69720/.

28. "MoveOn.Org Political Action: Democracy in Action," accessed February 2, 2011, http://pol.moveon.org/archive/.

29. "MoveOn.Org Political Action: Democracy in Action."

30. Jay David Bolter and Richard Grusin, *Remediation: Understanding New Media* (MIT Press, 2000), 208.

31. "MoveOn.Org: Join the MoveOn Media Corps," February 7, 2004, https://web.archive.org/web/20050216014239/http://www.moveon.org/mediacorps/.

32. Charles Musser, "Political Documentary.

33. Pariser, "E-mail from Move-On re: Fahrenheit 9/11," https://web.archive.org/web/20040618113457/http://www.moveon.org/pac/news/f911.html.

34. Terry Eastland, "Filibuster Again! And Again!," *Weekly Standard*, May 19, 2003, https://www.weeklystandard.com/terry-eastland-for-the-editors/filibuster-again-and-again.

35. MoveOn collected the full set of sources for the accusations on its website. See MoveOn.org, "MoveOn.Org Civic Action: Documentation for 'Fair and Balanced?,'" accessed February 8, 2011, http://civic.moveon.org/mediacorps//fairandbalanced_sources.html.

36. Boynton, "Guerrilla Documentary."

37. Ken Auletta, "Vox Fox: How Roger Ailes and Fox News Are Changing Cable News," *New Yorker*, May 19, 2003, www.newyorker.com/magazine/2003/05/26/vox-fox.

38. Ibid.

39. Jonathan S. Morris, "The Fox News Factor," *Harvard International Journal of Press/Politics* 10, no. 3 (Summer 2005): 56–79, https://doi.org/10.1177/1081180X05279264; Stefano DellaVigna and Ethan Kaplan, "The Fox News Effect: Media Bias and Voting*," *Quarterly Journal of Economics* 122, no. 3 (August 2007): 1187–1234, https://doi.org/10.1162/qjec.122.3.1187.

40. MoveOn.org, "MoveOn.Org Civic Action: Democracy in Action: Fox Watch Sign Up," accessed February 8, 2011, http://civic.moveon.org/mediacorps//foxwatch/fox_survey.html?id = 2135-1460238-.tP8ZxoYFX906v8Z5Hjbbg.

41. Boynton, "Guerrilla Documentary."

42. See MoveOn.org, ""MoveOn.Org Civic Action: Unfair and Unbalanced," accessed February 8, 2011, http://civic.moveon.org/fox//; and "MoveOn.Org Civic Action: Democracy in Action: Fox Comments," accessed February 8, 2011, http://civic.moveon.org/news//fox-comments.html.

43. "Press Release: The Republican News Channel," accessed February 8, 2011, https://web.archive.org/web/20050701202219/https://www.moveon.org/press/pdfs/foxparrotsbush.pdf.

44. Auletta, "Vox Fox."

45. "OUTFOXED: Take Action!," accessed February 9, 2011, http://www.outfoxed.org/ActivismPartners.php.

46. Boynton, "Guerrilla Documentary."

47. Ibid.

48. Seth Ackerman, "The Most Biased Name in News," August 2001, http://www.fair.org/index.php?page = 1067; Hamm, *New Blue Media*.

49. Julia Lesage, "Torture Documentaries," *Jumpcut: A Review of Contemporary Media* 51 (Spring 2009), https://www.ejumpcut.org/archive/jc51.2009/TortureDocumentaries/.

50. Jeff Howe, *Crowdsourcing: Why the Power of the Crowd Is Driving the Future of Business* (Random House, 2008); Jeff Howe, "The Rise of Crowdsourcing," *Wired*, June 2006, www.wired.com/wired/archive/14.06/crowds.html.

51. The phrase here is Irani's. See Lilly Irani, "Difference and Dependence among Digital Workers: The Case of Amazon Mechanical Turk," *South Atlantic Quarterly* 114, no. 1 (2015): 226. For additional theorization in this area, see also Tiziana Terranova, "Free Labor: Producing Culture for the Digital Economy," *Social Text* 18, no. 2 (2000): 33–58; and Trebor Scholz, ed., *Digital Labor: The Internet as Playground and Factory* (Routledge, 2012).

52. Alongside her collaborator M. Six Silberman, Irani has built one such platform, Turkopticon, that allows workers to review and rate the requesters who solicit work through Mechanical Turk. See Lilly Irani, "Amazon Mechanical Turk," in *The Blackwell Encyclopedia of Sociology*, ed. G. Ritzer (John Wiley and Sons, 2007); and Lilly C. Irani and M. Six Silberman, "Turkopticon: Interrupting Worker Invisibility in Amazon Mechanical Turk," in *Proceedings of the SIGCHI Conference on Human Factors in Computing Systems* (ACM, 2013), 611–20, https://doi.org/10.1145/2470654.2470742.

53. See Box Office Mojo, "Avatar." https://www.boxofficemojo.com/movies/?id=avatar.htm.

54. David Whiteman, "Out of the Theaters and into the Streets: A Coalition Model of the Political Impact of Documentary Film and Video," *Political Communication* 21, no. 1 (2004): 51, https://doi.org/10.1080/10584600490273263–1585.

55. Randy Kennedy, "A Screening with Stars but a Focus on Politics," *New York Times*, November 6, 2003, www.nytimes.com/2003/11/06/arts/a-screening-with-stars-but-a-focus-on-politics.html?scp%20=%20&sq%20=; Rick Marshall, "Party for the Right to Fight; Local Activists Gather to Screen Anti-Fox Film as Part of a Nationwide Day of House Parties," *Metroland* 27, no. 30 (July 22, 2004): 14.

56. As David Norman Rodowick asserts, "[H]ome theater has already overtaken commercial exhibition in popularity and economic importance." Rodowick, *The Virtual Life of Film* (Harvard University Press, 2007), 109.

57. Interestingly, although video posts were only semifrequent, the site's message boards regularly got posts from users, indicating that the site was a destination for people interested in the issue long after the film's release. Its tongue-in-cheek approach to the material runs closer to parody than the film version does, but the site also offered a special *FOX ATTACKS: Special Edition* version of the *Outfoxed* film that includes some of the viral videos created for the website alongside the film. The original site can still be accessed via the Wayback Machine at archive.org. As of 2019, the URL redirects to a YouTube video with the full original film, and the comments section contains material similar to the sort of debate that appeared on the original FoxAttacks website.

58. Steven Levitt and Stephen Dubner, *Freakonomics: A Rogue Economist Explores the Hidden Side of Everything* (Harper Perennial, 2009). See also Levitt and Dubner, *Freakonomics: The Hidden Side of Everything* (blog/website/podcast) http://freakonomics.com. The site initially functioned as a frequently updated blog, but since 2017 it seems to have shifted its attention to producing the radio show and podcast as a way of generating interest and attention in the books.

59. Boynton, "Guerrilla Documentary."

60. Jonathan Kahana, *Intelligence Work: The Politics of American Documentary* (Columbia University Press, 2008), 3.

61. Consider, for instance, the enormous popularity of Netflix's "Watch Instantly" streaming option.

62. Musser, "Political Documentary."

63. *If Mueller Is Fired, Here's What to Do,* accessed August 2, 2018, https://www.facebook.com/moveon/videos/10155343022170493/.

64. *Host an Event!,* 2007, accessed June 15, 2019, http://www.youtube.com/watch?v=tAyGUjIusRM&feature=youtube_gdata_player.

65. Paul Arthur, "Jargons of Authenticity: Three American Moments," in *Theorizing Documentary,* ed. Michael Renov (Routledge, 1993), 108–35.

66. Jane M. Gaines, "Political Mimesis," in *Collecting Visible Evidence,* ed. Jane M. Gaines and Michael Renov (University of Minnesota Press, 1999), 85.

67. Jane M. Gaines and Michael Renov, eds., *Collecting Visible Evidence* (University of Minnesota Press, 1999).

68. Ellen Jones, "Reel to Real: Can Documentaries Change the World? *The Guardian,* October 6, 2011, www.guardian.co.uk/film/2011/oct/06/documentaries-puma-creative-impact-award?CMP = EMCGT_071011&.

69. Ibid.

70. Malcolm Gladwell, "Small Change," *New Yorker,* October 4, 2010, http://www.newyorker.com/reporting/2010/10/04/101004fa_fact_gladwell?currentPage = all.

71. Ibid.

72. Salwa Ismail, "Civilities, Subjectivities and Collective Action: Preliminary Reflections in Light of the Egyptian Revolution," *Third World Quarterly* 32, no. 5 (2011): 989–95, https://doi.org/10.1080/01436597.2011.578976.

73. Helen Margetts, Peter John, Scott Hale, and Taha Yasseri, *Political Turbulence: How Social Media Shape Collective Action* (Princeton University Press, 2016).

74. Shirky, *Here Comes Everybody,* 106.

75. Evgeny Morozov, *The Net Delusion: The Dark Side of Internet Freedom* (PublicAffairs, 2011), xvi.

76. The argument here strongly echoes the argument that Wendy Chun puts forth in her first book, although she is primarily concerned in this early iteration with the individual rather than a larger social/organizational institution. See Wendy Hui Kyong Chun, *Control and Freedom: Power and Paranoia in the Age of Fiber Optics* (MIT Press, 2008).

77. Morozov, *The Net Delusion,* 309.

78. Shirky, *Here Comes Everybody.*

79. Manuel Castells, *The Internet Galaxy: Reflections on the Internet, Business, and Society* (Oxford University Press, 2002), 17–25.

80. Gladwell, "Small Change."

81. Ibid.

82. José van Dijck, *The Culture of Connectivity: A Critical History of Social Media,* (Oxford University Press, 2013), 32.

83. José van Dijck, "Opening Keynote: 'The Platform Society,'" accessed August 5, 2018, www.youtube.com/watch?v=-ypiiSQTNqo.

84. Jane M. Gaines, "The Production of Outrage: The Iraq War and the Radical Documentary Tradition," *Framework: The Journal of Cinema and Media* 48, no. 2 (Fall 2007): 36–55.

85. Ibid., 46.

86. Ibid., 49.

87. Lawrence Lessig, *Remix: Making Art and Commerce Thrive in the Hybrid Economy* (Penguin Books, 2009), 85.

88. Whiteman, "Out of the Theaters"; Barbara Abrash and David Whiteman, "The Uprising of '34: Filmmaking as Community Engagement," *Wide Angle* 21, no. 2 (March 1999): 87–99.

89. Whiteman, "Out of the Theaters."

4. "STATES OF EXCEPTION"

1. Linton Weeks, "Obama Orders Guantanamo Bay Prison Closure," NPR, January 22, 2009, www.npr.org/templates/story/story.php?storyId=99728679.

2. The cynical view of the delay, one with certain foundations, is that it was a product of the administration's desire not to close the base at all, or to close the base in name only by shifting its duties (and crimes) to other locations cloaked under other legal auspices (like the detention center at Bagram, Afghanistan). Glenn Greenwald, "An Emerging Progressive Consensus on Obama's Executive Power and Secrecy Abuses," *Salon,* April 13, 2009, www.salon.com/2009/04/13/obama_94/.

3. Ashley Parker, "Rand Paul Leads Filibuster of Brennan Nomination," *The Caucus* (blog), March 6, 2013, http://thecaucus.blogs.nytimes.com/2013/03/06/rand-paul-filibusters-brennan-nomination/.

4. The term here is drawn from the title, and argument, of Derek Gregory's essay in the collection that he coedited with Allan Pred. While Gregory explores both Guantánamo and Abu Ghraib as vanishing points, he activates the wider set of exceptional policies, practices, and fields of engagement that I would point to as well. See Derek Gregory, "Vanishing Points," in *Violent Geographies: Fear, Terror, and Political Violence,* ed. Derek Gregory and Allan Pred (Routledge, 2007), 205.

5. P. W. Singer, *Wired for War: The Robotics Revolution and Conflict in the 21st Century,* reprint (Penguin Books, 2009).

6. Timothy Lenoir and Luke Caldwell, *The Military-Entertainment Complex* (Harvard University Press, 2017).

7. By "represent," I'm referring here both to efforts to offer political and legal representation to the detainees of Guantánamo and to the efforts by media makers to figuratively represent the place on-screen—something that throughout the Bush era was notoriously difficult to do. It is the conflation of the political and the aesthetic (following Rancière in *The Politics of Aesthetics*) that sits at the heart of the chapter that follows.

8. The project's blog, http://gonegitmo.blogspot.com, recounts the various stages, sabotages, and setbacks the project went through before reaching its current state as well as the various conferences, events, and press coverage at or in which it has been discussed since.

9. Nonny de la Peña and Peggy Weil, "Gone Gitmo: ACLU Blog Post," *Gone Gitmo* (blog), January 11, 2008, https://gonegitmo.blogspot.com/2008/01/aclu-calls-11108-event.html.

10. "History of Second Life," Second Life Wiki, accessed December 16, 2018, http://wiki.secondlife.com/wiki/History_of_Second_Life.

11. The relationship between fiction film and the historical past is, of course, well-plowed ground. See, for example, Robert Rosenstone, *History on Film/Film on History* (Pearson, 2006).

12. Bill Nichols, *Blurred Boundaries: Questions of Meaning in Contemporary Culture* (Indiana University Press, 1994).

13. In her recent *Jump Cut* article "Torture Documentaries," Julia Lesage puts the issue of documentary's facts-cum-ethics thus: "Their films give information about the subject, indicate ways of dealing with the issues, invite an emotional response, and invoke an ethical stance. They offer a path to mastery over a complex topic, even if it is only a provisional mastery that becomes more nuanced and revised the more we consider other facts and other voices on the subject." Julia Lesage, "Torture Documentaries," *Jumpcut: A Review of Contemporary Media* 51 (Spring 2009), http://www.ejumpcut.org/currentissue/TortureDocumentaries/index.html.

14. Fiction film, of course, has its own, less overt politics, and the same could certainly be said of Second Life in general, but again, *Gone Gitmo*'s overt approach to politics puts it in the same oppositional stance to the rest of Second Life that documentary occupies in relation to the majority of fiction film.

15. Nadia Orenes, "The Camera Angle as an Expressive Resource and Narrative Booster in Video Games," *Gamasutra* (blog), January 15, 2018, www.gamasutra.com/blogs/NadiaOrenes/20180115/312999/The_camera_angle_as_an_expressive_resource_and_narrative_booster_in_video_games.php.

16. As reflected on the site's blog, the closing of Camp X-Ray and the construction of Camp Delta are both developments that took place during the site's construction. Nonny De La Peña, "Gone Gitmo: C-17, Camp Delta and a Visit from Gandhi," *Gone Gitmo* (blog), April 23, 2008, https://gonegitmo.blogspot.com/2008/04/c-17-camp-delta-and-visit-from-gandhi.html.

17. Ironically, there is nonetheless a kernel of photographic "truth" lurking in some of the images on the site in that its designers used digital images of the camp itself to create faithful virtual images in Second Life. For an account of replicating the actual razor wire used in the real camp, see Peggy Weil, "Gone Gitmo: Building Cages," *Gone Gitmo* (blog), June 7, 2007, https://gonegitmo.blogspot.com/2008/01/building-cages.html.

18. See, for example, the comments on the project's blog page at http://gonegitmo.blogspot.com.

19. Ben Fox, "Military Blocks Media Access to Guantanamo," *Washington Post,* June 15, 2006, www.washingtonpost.com/wp-dyn/content/article/2006/06/15/AR2006061501506.html.

20. Nonny de la Peña's later project *After Solitary* was an effort to achieve this same experience using VR technology. See Lilly Knoepp, "Forget Oculus Rift, Meet the Godmother

of VR," *Forbes,* accessed August 23, 2018, www.forbes.com/sites/lillyknoepp/2017/04/13/forget-oculus-rift-meet-the-godmother-of-vr/.

21. Peggy Weil, "Gone Gitmo: MMOMA Exhibition Diagram," *Gone Gitmo* (blog), July 12, 2013, http://gonegitmo.blogspot.com/2013/07/mmoma-exhibition-diagram.html; Peggy Weil, "Gone Gitmo: Images from Culture Project Event," *Gone Gitmo* (blog), June 8, 2010, http://gonegitmo.blogspot.com/2010/06/images-from-culture-project-event.html.

22. Ben Fino-Radin, "Art in the Age of Obsolescence: Rescuing an Artwork from Crumbling Technologies," MoMA, December 21, 2016, https://stories.moma.org/art-in-the-age-of-obsolescence-1272f1b9b92e; Jon Ippolito, "Ten Myths of Internet Art," *Leonardo* 35, no. 5 (October 2002): 485–98, https://doi.org/10.1162/002409402320774312; Frank Rose, "The Mission to Save Vanishing Internet Art," *New York Times,* October 21, 2016, www.nytimes.com/2016/10/23/arts/design/the-mission-to-save-vanishing-internet-art.html.

23. Leslie Jamison, "The Digital Ruins of a Forgotten Future," *The Atlantic,* December 2017, www.theatlantic.com/magazine/archive/2017/12/second-life-leslie-jamison/544149/.

24. For Agamben, this specific "state" is neither a historical development (indeed, its antecedents include the Roman *iustitium* and *auctoritas,* used in state funerals) nor the product of any specific form of constitution. In this sense, 9/11, Guantánamo and the Patriot Act aren't so much new developments as they are clear examples of the extreme limits to which state sovereignty and individual rights can be pushed under the right conditions. In his book *State of Exception,* Agamben himself makes the rather scandalous comparison between Hitler's "Decree for the Protection of the People" and the Patriot Act itself. What this persistence points to for Agamben is the impossibility of any outside to the "state of exception," which also nullifies any possibility of a future free of such things. (In essence, the rule always contains the possibility of its exception; the exception is the condition of possibility for the rule, or maybe even the act of ruling as such.) While this may be so, it nonetheless fails to account for the possibility that developments like the Patriot Act or Guantánamo can be reversed and closed, respectively—something that the forms of representation I'm addressing here specifically seek to achieve. The exception may always exist, but as a potentiality rather than an actuality—a difference that would have significance for those groups and individuals subjected to its application. See Agamben, *State of Exception* (University of Chicago Press, 2005); and his earlier text *Homo Sacer: Sovereign Power and Bare Life* (Stanford University Press, 1998), in which he more fully discusses the foundations of sovereignty.

25. Agamben, *State of Exception,* 50–51.

26. Derek Gregory, "The Black Flag: Guantánamo Bay and the Space of Exception," *Geografiska Annaler: Series B, Human Geography* 88, no. 4 (2006): 411.

27. Butler's primary goal in this text is primarily discursive in the sense that she is particularly concerned with demonstrating how the replacement of one form of discourse (considerations of why the United States was attacked on 9/11) with another (the political designation of detainees as "illegal enemy combatants") carries political, ethical, and moral stakes. Moreover, for my purposes here, the representation of the Other to the self, particularly in forms of media representation and their exclusion or censorship, is part of the way in which such discursive shifts are enacted. See Judith Butler, *Precarious Life: The Power of Mourning and Violence* (Verso, 2006).

28. The Uighurs I refer to here are Muslim Chinese separatists who went to Afghanistan to receive terrorist training in order to overthrow their secular government. They were rounded up after the invasion and ended up in Guantánamo. As proclaimed enemies of the state, they clearly could not be sent back to China, and yet they can't be legally detained as a threat to the United States (not to mention the political outrage the suggestion of resettling them in the United States sparked). Where, then, to send them? In an aid deal negotiated in 2009 by the Obama administration with the governments of Bermuda and Palau, the Uighurs will end up on one of the few island nations smaller than Cuba.

29. Carlos Williams, "Guantánamo Echoes U.S. 'Gunboat' Past / Anti-American Forces Use Navy Base as Rallying Symbol," Los Angeles Times, Sunday April 22, 2007. The original terms of the contract stipulated that the United States pay the Cuban government, in the form of a check, an annual sum of $4,085, which it has faithfully sent every year; but the checks have remained uncashed since 1960. Purportedly, Castro keeps the checks themselves in a drawer in his desk. Long before September 11, 2001, and the prison camp that it would eventually bring, US occupation of the base itself and the terms under which the United States has occupied the territory have been a point of contention for otherwise politically friendly countries in the region. To add to the irony, there is now a proposal by the Council on Hemispheric Relations to litigate the United States off the island by claiming that the original terms of the lease, which prohibited commercial operations of any sort, were broken by the military when it opened McDonald's, Starbucks, and Subway locations on the base in order to offer soldiers stationed there some of the amenities of home.

30. Amy Kaplan, "Prisoners and Rights: Guantánamo's Limbo Is Too Convenient," New York Times, November 24, 2003, www.nytimes.com/2003/11/24/opinion/24iht-edkaplan_ ed3_.html. See also Amy Kaplan, "Where Is Guantanamo?," American Quarterly 57, no. 3 (September 2005): 831–58, https://doi.org/10.1353/aq.2005.0048, Kaplan's longer article from which the New York Times article is drawn.

31. Kaplan, "Where Is Guantanamo?"

32. Ibid.

33. Taken from Second Life's own reporting, which is the only available source for traffic information. "The Second Life Economy in Q3 2011," Second Life Community, accessed June 17, 2019, https://community.secondlife.com/blogs/entry/288-the-second-life-economy-in-q3-2011/.

34. A great deal of Dreyfus's work has gone into working out what artificial intelligence systems (both those in existence and those in development) fail to achieve in terms of philosophical standards and expectations of "mind" and cognition, and why they fail to achieve them. See Hubert Dreyfus, What Computers Can't Do (Harper and Row, 1972); and his follow-up book, What Computers Still Can't Do (MIT Press, 1992).

35. See Hubert Dreyfus, "Virtual Embodiment: Myths of Meaning in Second Life," in On the Internet, 2nd ed. (Routledge, 2008). A phenomenologist of the first order, Dreyfus draws from both Martin Heidegger ("focal experience") and Maurice Merleau-Ponty (gesture and intercorporeality) for his discussion of what's most rewarding in real life—sources that are then supplemented with recent research in cognitive and neuropsychology (mirror neurons, social proxemics, etc.).

36. See Tom Boellstorff, Coming of Age in Second Life: An Anthropologist Explores the Virtually Human (Princeton University Press, 2008). Of particular use here to me was chap. 5,

"Personhood: The Self—The Life Course—Avatars and Alts—Embodiment—Gender and Race—Agency."

37. Boellstorff's most extreme example of this is the case of a user named "Pavia," who occupied Second Life as a female but real life as a male. In the course of her time in Second Life, Pavia came to identify far more with the female gender in real life, to the extent that she later came to identify as transgender. Boellstorff, *Coming of Age,* 138–39.

38. Jamison, "Digital Ruins."

39. See Jay David Bolter and Richard Grusin, *Remediation: Understanding New Media* (MIT Press, 2000), in which the authors argue that new visual media achieve their cultural significance precisely by paying homage to, rivaling, and refashioning such earlier media as perspective painting, photography, film, and television. They call this process of refashioning "remediation," and they note that earlier media have also often refashioned one another: photography remediated painting; film remediated stage production and photography; and television remediated film, vaudeville, and radio.

40. The specific irony of using the term "black sites" to refer to those that don't exist under legal and rhetorical terms, but must, by definition, exist in some physical space (and hence be open to some form of representation) has been extensively mined by the geographer Trevor Paglen in his explorations of the secret, internal workings of the CIA and related government agencies that seek to work under the veil of secrecy. Paglen has produced two relevant studies of such places and their by-products. The first, *Torture Taxi,* which he coauthored with A. C. Thompson, investigates CIA policies of extraordinary rendition by capitalizing on the fact that such secret flights still have to adhere to the exigencies of any given flight, including taking off and landing at airports, refueling, and so forth. The second, *Blank Spots on the Map,* is a study that documents "black sites"—from Area 51 in Nevada, to the elusive "Salt Pit" in Afghanistan, to even more secret locations whose names are unknown. See Trevor Paglen and A. C. Thompson, *Torture Taxi: On the Trail of the CIA's Rendition Flights* (Melville House, 2006); and A. C. Paglen, *Blank Spots on the Map: The Dark Geography of the Pentagon's Secret World* (Dutton, 2009).

41. Lenoir and Caldwell, *The Military-Entertainment Complex.*

42. Caren Kaplan, "Drone-O-Rama," in *Life in the Age of Drone Warfare,* ed. Lisa Parks and Caren Kaplan (Duke University Press, 2017); Lenoir and Caldwell, *The Military-Entertainment Complex.*

43. Kaplan, "Drone-O-Rama"; Lenoir and Caldwell, *The Military-Entertainment Complex;* Derek Gregory, "Dirty Dancing, Drones and Death in the Borderlands," in *Life in the Age of Drone Warfare,* ed. Lisa Parks and Caren Kaplan (Duke University Press, 2017).

44. Lisa Parks, "Vertical Mediation and the US Drone War in the Horn of Africa," in *Life in the Age of Drone Warfare,* ed. Lisa Parks and Caren Kaplan (Duke University Press, 2017); Brandon Bryant, "Letter from a Sensor Operator," in *Life in the Age of Drone Warfare,* ed. Lisa Parks and Caren Kaplan (Duke University Press, 2017).

45. Lee Grieveson and Haidee Wasson, "The Military's Cinema Complex," in *Cinema's Military Industrial Complex,* ed. Haidee Wasson and Lee Grieveson (University of California Press, 2018), 2–4.

46. Ed Halter, *From Sun Tzu to Xbox: War and Video Games* (PublicAffairs, 2006); Ian Bogost, *How to Do Things with Videogames* (University of Minnesota Press, 2011); Alexander R. Galloway, "Social Realism in Gaming," *Game Studies* 4, no. 1 (2004): 23–26;

Nina B. Huntemann and Matthew Thomas Payne, eds., *Joystick Soldiers: The Politics of Play in Military Video Games* (Routledge, 2009).

47. James Der Derian, *Virtuous War: Mapping the Military-Industrial-Media-Entertainment Network*, 2nd ed. (Routledge, 2009); Roger Stahl, *Militainment, Inc.: War, Media, and Popular Culture* (Routledge, 2009).

48. Huntemann and Payne, *Joystick Soldiers*.

49. Lenoir and Caldwell, *The Military-Entertainment Complex*.

50. Ian Bogost, *Persuasive Games: The Expressive Power of Videogames* (MIT Press, 2010).

51. Huntemann and Payne, *Joystick Soldiers*.

52. "Marine Corps Base Camp Pendleton > Staff & Agencies > Assistant Chief of Staff G-3/5 > Training Support Division > MOUT Facilities > IIT," accessed June 17, 2019, www.pendleton.marines.mil/Staff-Agencies/Assistant-Chief-of-Staff-G-3-5/Training-Support-Division/MOUT-Facilities/IIT/.

53. Noah Tsika, "From Wartime Instruction to Superpower Cinema," in *Cinema's Military Industrial Complex*, ed. Haidee Wasson and Lee Grieveson (University of California Press, 2018).

54. Douglas Cunningham, "Imaging/Imagining Air Force Identity: 'Hap' Arnold, Warner Bros., and the Formation of the USAAF First Motion Picture Unit," *Moving Image* 5, no. 1 (2005): 95–124; Douglas A. Cunningham, *Imagining Air Force Identity: Masculinity, Aeriality, and the Films of the U.S. Army Air Forces First Motion Picture Unit* (PhD diss., University of California, Berkeley, 2009), http://catalog.hathitrust.org/api/volumes/oclc/641611265.html.

55. Kaia Scott, "Managing the Trauma of Labor: Military Psychiatric Cinema in World War II," in *Cinema's Military Industrial Complex*, ed. Haidee Wasson and Lee Grieveson (University of California Press, 2018).

56. *Loading Screen Series—America's Army*, YouTube, Loading Screen Series (America's Army), accessed December 18, 2018, https://www.youtube.com/watch?v=IWEJWGZWono.

57. Bogost, *Persuasive Games*.

58. Mark J. P. Wolf, *Building Imaginary Worlds: The Theory and History of Subcreation*, (Routledge, 2012).

59. Galloway, "Social Realism in Gaming."

60. Lenoir and Caldwell, *The Military-Entertainment Complex*.

61. Galloway, "Social Realism in Gaming."

62. Nick Dyer-Witheford and Greig de Peuter, *Games of Empire: Global Capitalism and Video Games* (University of Minnesota Press, 2009), 112.

63. Halter, *From Sun Tzu to Xbox*.

64. *The Last Starfighter*, directed by Nick Castle (1984); *War Games*, directed by John Badham (1983); *Cloak & Dagger*, directed by Richard Franklin (1984).

65. Halter, *From Sun Tzu to Xbox*, 118.

66. Singer, *Wired for War*.

67. Gregoire Chamayou, *A Theory of the Drone* (New Press, 2015), chap. 3, esp. 30–35.

68. In general, media effects is the study of the impact of media on individual and social behavior. Priming effects refers to the (controversial) theory that exposure to a certain

behavior pattern in a media context will "prime" the affected individual to repeat that behavior in real life. For a discussion of priming effects, see Leonard Berkowitz, Ronald Corwin, and Mark Heironimus, "Film Violence and Subsequent Aggressive Tendencies," *Public Opinion Quarterly* 27, no. 2 (Summer1963): 217–29, https://doi.org/10.1086/267162; and Elihu Katz, "Communications Research since Lazarsfeld," *Public Opinion Quarterly* 51, no. 4, pt. 2 (Winter 1987): S25–45, https://doi.org/10.1093/poq/51.4_PART_2.S25.

69. Peter Asaro, "The Labor of Surveillance and Bureaucratized Killing," in *Life in the Age of Drone Warfare,* ed. Lisa Parks and Caren Kaplan (Duke University Press, 2017), 308.

70. Parks, "Vertical Mediation," 135.

71. Charles Sanders Peirce, *Collected Papers of Charles Sanders Peirce* (Harvard University Press, 1974), 200.

72. Bill Nichols, *Representing Reality: Issues and Concepts in Documentary* (Indiana University Press, 1991).

73. Thomas Stubblefield, "In Pursuit of Other Networks," in *Life in the Age of Drone Warfare,* ed. Lisa Parks and Caren Kaplan (Duke University Press, 2017), 206.

74. Jefferson Morely, "Boredom, Terror, Deadly Mistakes: Secrets of the New Drone War," *Salon,* April 3, 2012, www.salon.com/2012/04/03/boredom_terror_deadly_mistakes_secrets_of_the_new_drone_war/.

75. Nicola Abé, "Dreams in Infrared: An Unpopular Job," *Spiegel Online,* December 14, 2012, www.spiegel.de/international/world/pain-continues-after-war-for-american-drone-pilot-a-872726-2.html.

76. Derek Gregory, "From a View to a Kill: Drones and Late Modern War," *Theory, Culture & Society* 28, nos. 7–8 (December 2011): 188–215, https://doi.org/10.1177/0263276411423027.

77. Derek Gregory, "Dirty Dancing, Drones and Death in the Borderlands," in *Life in the Age of Drone Warfare,* ed. Lisa Parks and Caren Kaplan (Duke University Press, 2017).

78. Asaro, "The Labor of Surveillance."

79. Elisabeth Bumiller, "Drone Pilots, Waiting for a Kill Shot 7,000 Miles Away," *New York Times,* July 29, 2012, www.nytimes.com/2012/07/30/us/drone-pilots-waiting-for-a-kill-shot-7000-miles-away.html.

80. Christiane Paul, "Augmented Realities: Digital Art in the Public Space," in *A Companion to Public Art,* ed. Cher Krause Knight and Harriet F. Senie (John Wiley & Sons, 2016), 205.

81. Joseph DeLappe, "Gandhi's 'Salt March to Dandi' in Second Life," *Reenactment: The Salt Satyagraha Online* (blog), March 5, 2008, https://saltmarchsecondlife.wordpress.com/about/.

82. Joseph DeLappe, "Dead-in-Iraq: Performance/Memorial/Protest," *TDR/The Drama Review* 52, no. 1 (Spring 2008): 2–3.

83. These soldiers are often consulted on their experiences for future iterations of the game. Oddly enough, the game goes one step further and produces collectible action figures of these soldiers, thereby allowing very young children to engage in the play of war. See Jacqueline M. Hames, "America's Army—Army Values & Plenty of Action," U.S. Army, accessed April 16, 2013, http://www.army.mil/article/26405/.

84. David Axe, "How to Prevent Drone Pilot PTSD: Blame the 'Bot" Wired.Com, June 7, 2012, www.wired.com/dangerroom/2012/06/drone-pilot-ptsd/.

5. TECHNOLOGY, TRANSPARENCY, AND THE DIGITAL PRESIDENCY

The quotations in the epigraph are from, respectively, Louis Brandeis, *Other People's Money: And How the Bankers Use It* (F. A. Stokes, 1914), 92; and Edward Tufte, *Beautiful Evidence* (Graphics Press, 2006), 9.

1. Ed O'Keefe, "Obama Finally Accepts His Transparency Award . . . behind Closed Doors," March 31, 2011, *Washington Post,* www.washingtonpost.com/blogs/federal-eye/post/obama-finally-accepts-his-transparency-award-behind-closed-doors/2011/03/31/AFRplO9B_blog.html.

2. "About Sunshine Week," accessed May 12, 2011, https://web.archive.org/web/2011 0416022233/www.sunshineweek.org/About.aspx The current website can be found here: http://sunshineweek.org.

3. John Tagg, *The Burden of Representation: Essays on Photographies and Histories* (University of Minnesota Press, 1993), 20; and Richard Hofstadter, *The Age of Reform* (Vintage, 1960), 19.

4. "IBM Builds a Smarter Planet," IBM, April 8, 2011, www.ibm.com/smarterplanet/us/en/?ca = v_smarterplanet; *Why Data Matters: IBM, Let's Build a Smarter Planet,* 2011, www.youtube.com/watch?v=pkBwB8ofcXI.

5. Paul Farhi, "CNN Hits the Wall for the Election," Washingtonpost.com," February 5, 2008, www.washingtonpost.com/wp-dyn/content/article/2008/02/04/AR2008020402796.html; Shannon Doyne, Holly Epstein Ojalvo, and Katherine Schulten, "Data Visualized: More on Teaching with Infographics," *The Learning Network: Teaching and Learning with the New York Times* (blog), April 8, 2011, http://learning.blogs.nytimes.com/2011/04/08/data-visualized-more-on-teaching-with-infographics/?partner = rss&emc = rss.

6. For the way large datasets become stories, see, for example, Simon Rogers, "Data Journalism Broken Down: What We Do to the Data before You See It ," *The Guardian,* April 7, 2011, https://www.theguardian.com/news/datablog/2011/apr/07/data-journalism-workflow. For background on the field, see Geoff McGhee, *Journalism in the Age of Data,* https://datajournalism.stanford.edu. (McGhee prepared this online report through a John S. Knight grant at Stanford University.)

7. Edward Tufte, *The Visual Display of Quantitative Information,* 2nd ed. (Graphics Press, 2001), 2.

8. The literature here is vast. For a discussion of sensor networks, see Stacey Higginbotham, "Sensor Networks Top Social Networks for Big Data," *GigaOm* (blog), September 13, 2010, http://gigaom.com/cloud/sensor-networks-top-social-networks-for-big-data-2/; and Gregory Pfister, *In Search of Clusters,* 2nd ed. (Prentice Hall, 1997). For a more general history of the recent development of data visualization, including what the authors define as the fifteen-year "foundational period of the field," see Stuart K. Card, Jock D. Mackinlay, and Ben Shneiderman, *Readings in Information Visualization: Using Vision to Think* (Morgan Kaufmann, 1999), xiii.

9. Tufte, *Beautiful Evidence,* 10; Card, Mackinlay, and Shneiderman, *Readings in Information Visualization,* xiii.

10. See Jeffrey Heer's work and the general mission of his lab at the University of Washington: http://idl.cs.washington.edu/.

11. Michael Friendly, "The Golden Age of Statistical Graphics," *Statistical Science* 23, no. 4 (2008): 502–35.

12. Ibid., 532.

13. Ian Hacking, *The Taming of Chance* (Cambridge University Press, 1990), 1.

14. Friendly, "The Golden Age of Statistical Graphics," 505. Various chapters of Hacking, *The Taming of Chance,* also focus on one or another of these figures.

15. Hacking, *The Taming of Chance,* 3.

16. Michel Foucault, *Security, Territory, Population: Lectures at the College de France, 1977–1978* (Picador, 2009), 274.

17. Robin Kelsey and Blake Stimson, *The Meaning of Photography* (Clark Art Institute, 2008), 3–4.

18. Alan Trachtenberg, ed., *Classic Essays on Photography* (Leete's Island Books, 1980), 20–24.

19. Kelsey and Stimson, *The Meaning of Photography,* xii.

20. Lorraine Daston and Peter Galison, *Objectivity* (Zone Books, 2010).

21. Kelsey and Stimson, *The Meaning of Photography,* xv.

22. Trachtenberg, *Classic Essays on Photography,* 74.

23. Ibid., 81; Friendly, "Golden Age of Statistical Graphics," 505.

24. Lisa Cartwright, "'Experiments of Destruction': Cinematic Inscriptions of Physiology," *Representations,* no. 40 (1992): 129–52, https://doi.org/10.2307/2928742.

25. Kelsey and Stimson, *The Meaning of Photography,* xv.

26. Ibid., xi.

27. Tagg, *The Burden of Representation.*

28. Mike Davis, *City of Quartz: Excavating the Future in Los Angeles* (Verso, 2006); Lisa Parks, *Cultures in Orbit: Satellites and the Televisual* (Duke University Press, 2005).

29. Trachtenberg, *Classic Essays on Photography,* 109.

30. Paula Rabinowitz, *They Must Be Represented* (Verso, 1994); Tagg, *The Burden of Representation.*

31. Lewis Hine, quoted in Trachtenberg, *Classic Essays on Photography,* 112.

32. Jonathan Kahana, *Intelligence Work: The Politics of American Documentary* (Columbia University Press, 2008).

33. Ibid., 33.

34. K. Stoker and B. L Rawlins, "The 'Light' of Publicity in the Progressive Era: From Searchlight to Flashlight," *Journalism History* 30, no. 4 (2005): 177–88.

35. Hacking, *The Taming of Chance,* chap. 3.

36. Martin Campbell-Kelly, William Aspray, Nathan Ensmenger, and Jeffrey R. Yost, *Computer: A History of the Information Machine,* 3rd ed. (Westview Press, 2013), 13–18, 119–43.

37. Tufte, *Beautiful Evidence,* 10.

38. Martin Jay, *Downcast Eyes: The Denigration of Vision in Twentieth-Century French Thought* (University of California Press, 1994), chaps. 1–2; W. J. T. Mitchell, *Iconology: Image, Text, Ideology* (University of Chicago Press, 1987); W. J. T. Mitchell, *Picture Theory: Essays on Verbal and Visual Representation* (University of Chicago Press, 1995).

39. Jay, *Downcast Eyes,* 3–15.

40. Card, Mackinlay, and Shneiderman, *Readings in Information Visualization,* 1.

41. For Shneiderman's lab website, see www.cs.umd.edu/hcil/.

42. Card, Mackinlay, and Shneiderman, *Readings in Information Visualization*, 1.

43. Fran Smith, "Intelligent Designs: When Information Needs to Be Communicated, Edward Tufte Demands Both Truth and Beauty," *Stanford Magazine*, March/April 2007, https://stanfordmag.org/contents/intelligent-designs.

44. Tufte, *Beautiful Evidence*, 9.

45. Edward R. Tufte, *The Cognitive Style of PowerPoint: Pitching Out Corrupts Within*, 2nd ed. (Graphics Press, 2006).

46. Jeff Zeleny, "Lose the BlackBerry? Yes He Can, Maybe," *New York Times*, November 16, 2008, www.nytimes.com/2008/11/16/us/politics/16blackberry.html; Ann Sanner, "Obama's YouTube Address," *New York Daily News*, November 14, 2008, www.nydailynews.com/news/politics/obama-youtube-address-article-1.334621.

47. Samuel Greengard, "The First Internet President," *Communications of the ACM* 52, no. 2 (February 2009): 16–18; Sandford Borins, "From Online Candidate to Online President," *International Journal of Public Administration* 32, no. 9 (2009): 753.

48. "President Obama Embraces Openness on Day One, as Urged by the National Security Archive and a Coalition of More Than 60 Organizations," National Security Archive, January 21, 2009, https://nsarchive2.gwu.edu/news/20090121/.

49. Micah L. Sifry, *WikiLeaks and the Age of Transparency* (Counterpoint, 2011), 87–107.

50. Judie Attard, Fabrizio Orlandi, Simon Scerri, and Sören Auer, "A Systematic Review of Open Government Data Initiatives," *Government Information Quarterly* 32, no. 4 (October 2015): 399–418, https://doi.org/10.1016/j.giq.2015.07.006.

51. William Davies, *The Limits of Neoliberalism: Authority, Sovereignty and the Logic of Competition* (SAGE Publications, 2016); Rob Kitchin, "The Real-Time City? Big Data and Smart Urbanism," *GeoJournal* 79, no. 1 (February 2014): 1–14, https://doi.org/10.1007/s10708-013-9516-8.

52. Ironically, Kundra began his career in public service by interviewing for and receiving an IT job for Arlington County, Virginia, on the same day as the September 11, 2001, terrorist attacks.

53. Karim R. Lakhani, Robert D. Austin, and Yumi Yi, "Data.gov," Harvard Business School Case no. 610-075, May 2010, 4–6.

54. "Data Catalog," accessed May 14, 2011, http://data.dc.gov/.

55. Ibid.

56. "District of Columbia's Data Feeds Wins Innovations Award," Ash Center, Harvard Kennedy School, September 15, 2009, https://ash.harvard.edu/news/district-columbias-data-feeds-wins-innovations-award.

57. "Your Money at Work." Retrieved via the Wayback Machine Internet archive, https://archive.org/web/.

58. Ibid.

59. Jane Gaines and Michael Renov, *Collecting Visible Evidence* (University of Minnesota Press, 1999), 51–52.

60. See, for example, the cover of the November 24, 2008, issue of *Time* magazine, which featured a photocollage that placed Obama's smiling face on a famous photograph of FDR driving, with the headline "The New New Deal"; or Steve Lohr, "F.D.R.'s Example

Offers Lessons for Obama," *New York Times,* January 27, 2009, ww.nytimes.com/2009/01/27/business/economy/27fdr.html.

61. Paul Arthur, "Jargons of Authenticity (Three American Moments)," in *Theorizing Documentary,* ed. Michael Renov (Routledge, 1993), 116–17.

62. For a discussion of the use of classical rhetorical forms in documentary film, see Bill Nichols, *Representing Reality* (Indiana University Press, 1991), 134–41.

63. Dan Nguyen, "Blog like Edward Tufte: High Art WordPress Theme," *Danwin. Com* (blog), May 18, 2011, http://danwin.com/2011/05/blog-like-edward-tufte-high-art-wordpress-theme/; David Smith, "The Stimulus, Mapped," *R-Bloggers* (blog), October 29, 2010, www.r-bloggers.com/the-stimulus-mapped/.

64. Lakhani, Austin, and Yi, "Data.gov," 13–14.

65. Chris Vein, "Open Government Plans' Anniversary Is a Testament to Hard Work at Agencies," *White House Open Government Initiative* (blog), April 7, 2011, www.whitehouse.gov/blog/2011/04/07/open-government-plans-anniversary-testament-hard-work-agencies.

66. "Transparency and Open Government: Memorandum for the Heads of Executive Departments and Agencies," January 21, 2009, https://obamawhitehouse.archives.gov/the-press-office/transparency-and-open-government.

67. "Open Government Directive," December 8, 2009, https://obamawhitehouse.archives.gov/open/documents/open-government-directive.

68. Ibid.

69. Vein, "Open Government Plans' Anniversary."

70. Initially the site hosted the community forums on the data.gov website, but eventually this list was moved to Stack Exchange. See https://opendata.stackexchange.com.

71. Eliot Van Buskirk, "Sneak Peek: Obama Administration's Redesigned Data.gov," *Epicenter Wired.com* (blog), May 19, 2010, www.wired.com/epicenter/2010/05/sneak-peek-the-obama-administrations-redesigned-datagov/all/1.

72. The current apps that use the data.gov data sets can be browsed at https://www.data.gov/applications. Healthy Hive and ZocDoc are both health-care applications. The data set on which Airport Status Service was based can be found at https://catalog.data.gov/dataset/airport-status-web-service-9b3a2.

73. Lakhani, Austin, and Yi, "Data.gov."

74. Ibid.

75. Brian Winston, "The Documentary Film as Scientific Inscription," in *Theorizing Documentary,* ed. Michael Renov (Routledge, 1993), 37–57, http://eprints.lincoln.ac.uk/12108/.

76. Lakhani, Austin, and Yi, "Data.gov," 11.

77. Van Buskirk, "Sneak Peek."

78. Danny Vinik, "What Happened to Trump's War on Data?," *Politico,* July 25, 2017, www.politico.com/agenda/story/2017/07/25/what-happened-trump-war-data-000481.

79. Wendy Chun, "The Enduring Ephemeral, or the Future Is a Memory," *Critical Inquiry* 35, no. 1 (Autumn 2008): 148–71.

80. Quoted in Aliya Sternstein, "Tracing Transparency," *Government Executive* 43, no. 1 (January 2011): 2.

81. Quoted in ibid.

82. Noah Feldman, "In Defense of Secrecy," *New York Times,* February 15, 2009, www.nytimes.com/2009/02/15/magazine/15wwln_lede-t.html.

83. Quoted in "Sunshine and Shadows: The Clear Obama Message for Freedom of Information Meets Mixed Results," National Security Archive, March 15, 2010, https://nsarchive2.gwu.edu/NSAEBB/NSAEBB308/index.htm.

84. Beth Noveck, "Open Government Plans: A Tour of the Horizon," April 8, 2010, https://obamawhitehouse.archives.gov/blog/2010/04/08/open-government-plans-a-tour-horizon.

85. Ibid.

86. Robert Booth, Heather Brooke, and Steven Morris, "WikiLeaks Cables: Bradley Manning Faces 52 Years in Jail," *The Guardian,* November 30, 2010, www.guardian.co.uk/world/2010/nov/30/wikileaks-cables-bradley-manning?intcmp = 239.

87. Daniel Domscheit-Berg, *Inside WikiLeaks: My Time with Julian Assange at the World's Most Dangerous Website* (Crown, 2011).

88. As of the time of this writing, the exact role that Wikileaks played in the 2016 US presidential election and its surprising winner remains unknown, although it is at least clear that Assange timed the release of the DNC and Clinton campaign emails to be most damaging to Hillary Clinton. Assange released a half-hearted attempt to justify these actions in a tweet the day of the election; see https://wikileaks.org/Assange-Statement-on-the-US-Election.html.

89. Domscheit-Berg, *Inside WikiLeaks,* 8, 16–42.

90. Raffi Khatchadourian, "No Secrets," *New Yorker,* June 7, 2010, www.newyorker.com/reporting/2010/06/07/100607fa_fact_khatchadourian?currentPage = all.

91. Bill Keller, "Dealing with Assange and the WikiLeaks Secrets," *New York Times,* January 26, 2011, www.nytimes.com/2011/01/30/magazine/30Wikileaks-t.html?_r = 3&pagewanted = all.

92. Steven Colbert, "Exclusive—Julian Assange Extended Interview," *The Colbert Report* (Comedy Central, April 12, 2010), www.cc.com/video-clips/5mdm7i/the-colbert-report-exclusive---julian-assange-extended-interview.

93. Domscheit-Berg, *Inside WikiLeaks;* Colbert, "Julian Assange Extended Interview."

94. For a discussion of Grierson's use of the term and a critique, see Brian Winston, *Claiming the Real: Documentary: Grierson and Beyond,* 2nd ed. (British Film Institute, 2009).

95. The full quote is taken from Orwell's essay "Politics and the English Language" and reads: "Political language is designed to make lies sound truthful and murder respectable, and to give the appearance of solidity to pure wind."

96. Kahana, *Intelligence Work.*

97. Khatchadourian, "No Secrets."

98. See "Amy Goodman Reports on 'Collateral Murder' WikiLeaks Video," https://www.youtube.com/watch?v=7au7rMNY19I.

99. Ulrich Keller, *The Ultimate Spectacle: A Visual History of the Crimean War* (Gordon and Breach, 2001).

100. Sifry, *WikiLeaks.*

101. Consider, for example, the thoughts of Karl Friedrich Bahrdt, Friedrich Karl von Moser, and Johann Gottlieb Ficthe in *What Is Enlightenment?: Eighteenth-Century Answers and Twentieth-Century Question,* ed. James Schmidt (University of California Press, 1996).

102. Slavoj Žižek, "Good Manners in the Age of WikiLeaks," *London Review of Books,* January 20, 2011, www.lrb.co.uk/v33/n02/slavoj-zizek/good-manners-in-the-age-of-wikileaks.

103. Ibid.

104. Judith Butler, "Torture and the Ethics of Photography," *Environment and Planning D: Society and Space* 25, no. 6 (2007): 964, https://doi.org/10.1068/d2506jb.

105. Judith Butler, *Precarious Life: The Power of Mourning and Violence* (Verso, 2006).

106. Kevin A. Smith, "Review: Notice the Media at the Tip of the Spear," *Michigan Law Review* 102, no. 6 (May 1, 2004): 1329–72, https://doi.org/10.2307/4141948; Bill Katovsky and Timothy Carlson, *Embedded: The Media at War in Iraq* (Globe Pequot, 2004).

107. Katovsky and Carlson, *Embedded,* xi, 419.

108. The diplomatic cables seemed to offer more in the way of embarrassment than revelation and managed to tarnish the already fragile image of WikiLeaks as an organization bent on seeking the truth over public glory, and as Bill Keller notes, they offer a "very different kind of treasure" from the material related directly to the wars. See Keller, "Dealing with Assange."

109. "Secret Dispatches from the War in Iraq ," *New York Times,* accessed August 23, 2011, https://archive.nytimes.com/www.nytimes.com/interactive/world/iraq-war-logs.html#report/67845FF9-77DF-4982-98E9-A6951B54D025.

110. "Afghanistan War Logs: How the Guardian Got the Story" *The Guardian,* July 25, 2010, http://www.guardian.co.uk/world/2010/jul/25/afghanistan-war-logs-explained-video; "Piecing Together the Reports, and Deciding What to Publish," *New York Times,* July 25, 2010, www.nytimes.com/2010/07/26/world/26editors-note.html.

111. "Afghanistan: The War Logs," July 25, 2010, www.guardian.co.uk/world/the-war-logs.

112. C.J. Chivers, Carlotta Gall, Andrew W. Lehren, Mark Mazzetti, Jane Perlez, and Eric Schmitt, "View Is Bleaker Than Official Portrayal of War in Afghanistan," *New York Times,* July 25, 2010, www.nytimes.com/2010/07/26/world/asia/26warlogs.html. See also https://archive.nytimes.com/www.nytimes.com/interactive/world/war-logs.html?_r=0.

113. "Piecing Together the Reports, and Deciding What to Publish," *New York Times,* July 25, 2010, www.nytimes.com/2010/07/26/world/26editors-note.html.

114. Sifry, *WikiLeaks.*

115. Geert Lovink and Patrice Riemens, "Ten Theses on Wikileaks," Institute of Network Cultures, accessed June 18, 2019, http://networkcultures.org/geert/2010/08/30/ten-theses-on-wikileaks/.

116. Ibid.

117. Ibid.

6. POST-TRUTH POLITICS

1. This was even an opinion widely subscribed to by the Trump campaign team, including Trump himself. See Ben Schreckinger, "'He Thought He Was Going to Lose': Inside Donald Trump's Election Night War Room," *GQ,* November 7, 2017, www.gq.com/story/inside-donald-trumps-election-night-war-room.

2. This of course was the title of Clinton's book, which appeared less than a year after her loss to Donald Trump. For Clinton, the title is more of a declaration than a question, but in the days and weeks immediately after Trump's victory, suffice it to say that this was very much a question. Hillary Rodham Clinton, *What Happened* (Simon and Schuster, 2017).

3. Jesse Washington, "African-Americans See Painful Truths in Trump Victory," *The Undefeated* (blog), November 10, 2016, https://theundefeated.com/features/african-americans-see-painful-truths-in-trump-victory/.

4. Greg Sargent, "Why Did Trump Win? New Research by Democrats Offers a Worrisome Answer.," *Washington Post*, May 1, 2017, www.washingtonpost.com/blogs/plum-line/wp/2017/05/01/why-did-trump-win-new-research-by-democrats-offers-a-worrisome-answer/.

5. D. W. Pine, "Donald Trump Truth: Behind 'Is Truth Dead?,'" *Time*, April 3, 2017, http://time.com/4709920/donald-trump-truth-time-cover/.

6. Sasha Issenberg, *The Victory Lab: The Secret Science of Winning Campaigns* (Crown, 2012).

7. "Remarks on Internet Freedom," U.S. Department of State, January 21, 2010, https://2009-2017.state.gov/secretary/20092013clinton/rm/2010/01/135519.htm. This was a fairly regular refrain of the State Department itself during the tenure of Hillary Clinton as Secretary of State. See, for example, her speech cited in this note.

8. William Finnegan, "Donald Trump and the 'Amazing' Alex Jones," *New Yorker,* June 23, 2016, www.newyorker.com/news/daily-comment/donald-trump-and-the-amazing-alex-jones.

9. McKay Coppins, "How the Left Lost Its Mind," *The Atlantic,* July 2, 2017, www.theatlantic.com/politics/archive/2017/07/liberal-fever-swamps/530736/.

10. Within the fields of sociology, psychology, and cultural studies, questions have focused on the extent to which conspiracy thinking might be the product of some sort of neurosis or psychosis on the part of individuals and groups. The prevalence of conspiracy theories in cultures around the world for much of recorded history, however, seems to indicate that some general dimension of human cognition or recurring social pattern produces it, or that it in fact testifies to the presence of a vast global conspiracy. Barring from consideration the latter, these scholars explore the former.

Good summaries of the scholarship around conspiracy theories can be found in both Cass R. Sunstein and Adrian Vermeule, "Conspiracy Theories: Causes and Cures*," *Journal of Political Philosophy* 17, no. 2 (June 2009): 202–27, https://doi.org/10.1111/j.1467-9760.2008.00325.x; and Jovan Byford, *Conspiracy Theories: A Critical Introduction* (Palgrave Macmillan, 2011). For a good discussion of the neuropsychology research. see Rob Brotherton, *Suspicious Minds: Why We Believe Conspiracy Theories* (Bloomsbury Sigma, 2015).

11. Christopher Hodapp and Alice Von Kannon, *Conspiracy Theories & Secret Societies for Dummies* (Wiley, 2008).

12. Peter Knight, *Conspiracy Theories in American History: An Encyclopedia* (ABC-CLIO, 2003).

13. Popper's original essay and the set of different articles and arguments that it stimulated were brilliantly collected in a volume by David Coady. See Karl R. Popper, "The Conspiracy Theory of Society," in *Conspiracy Theories: The Philosophical Debate,* ed. David Coady (Ashgate, 2006).

14. Charles Pigden, "Popper Revisited, or What Is Wrong with Conspiracy Theories?," in *Conspiracy Theories: The Philosophical Debate,* ed. David Coady (Ashgate, 2006).

15. David Coady, "An Introduction to the Philosophical Debate about Conspiracy Theories," in *Conspiracy Theories: The Philosophical Debate,* ed. David Coady (Ashgate, 2006).

16. Brian L. Keeley, "Of Conspiracy Theories," in *Conspiracy Theories: The Philosophical Debate,* ed. David Coady (Ashgate, 2006), 57.

17. Richard Hofstadter, "The Paranoid Style in American Politics," *Harper's Magazine,* November 1964, https://harpers.org/archive/1964/11/the-paranoid-style-in-american-politics/; Richard Hofstadter, *The Paranoid Style in American Politics: And Other Essays* (Knopf, 1965).

18. Mark Fenster, *Conspiracy Theories: Secrecy and Power in American Culture,* 2nd ed. (University of Minnesota Press, 2008). In this book, Fenster tackles conspiracy theories both vast and complex, including the various factions of the 9/11 "truther" movement as well as the more casual fan communities that appeared around Dan Brown's megaselling *The Da Vinci Code.*

19. Fredric Jameson, *The Geopolitical Aesthetic: Cinema and Space in the World System* (Indiana University Press, 1995).

20. The literature on the Zapruder film is vast and, as I am arguing, places one directly in the labyrinth of Kennedy assassination lore. Peter Knight offers a good history of the film and its handling and reception post-assassination, including its inclusion in *Life* magazine. Zapruder's daughter, Alexandra Zapruder, has a personal history. Øyvind Vågnes's *Zaprudered* has an excellent discussion of the film's role within popular and visual culture. And James Fetzer's text offers an interesting example of the role of the film within conspiracy circles. See Peter Knight, *The Kennedy Assassination* (University Press of Mississippi, 2007); Alexandra Zapruder, *Twenty-Six Seconds: A Personal History of the Zapruder Film* (Grand Central Publishing, 2016); Øyvind Vågnes, *Zaprudered: The Kennedy Assassination Film in Visual Culture* (University of Texas Press, 2012); and James H. Fetzer, *The Great Zapruder Film Hoax: Deceit and Deception in the Death of JFK* (Open Court, 2013).

21. Stella Bruzzi, *New Documentary: A Critical Introduction* Routledge, 2011), 20–21.

22. Fenster, *Conspiracy Theories,* 252–63.

23. Mary Ann Doane, "Information, Crisis, Catastrophe," in *Logics of Television: Essays in Cultural Criticism,* ed. Patricia Mellenkamp (Indiana University Press, 1990), 222–39.

24. Fenster, *Conspiracy Theories.* This is where Fenster's work around the narrative structures of conspiracy theories provides a critical understanding of the form.

25. Peter Knight, "Outrageous Conspiracy Theories: Popular and Official Responses to 9/11 in Germany and the United States," *New German Critique,* no. 103 (Winter 2008): 165–93.

26. Henry Jenkins, *Convergence Culture: Where Old and New Media Collide* (NYU Press, 2006).

27. The difference of course is that consensus on climate change, and the connection between smoking and cancer before it, are the result of conscious, organized efforts by specific groups with a vested interest in changing public opinion. For a brilliant account of this, see Naomi Oreskes and Erik M Conway, *Merchants of Doubt: How a Handful of Scientists Obscured the Truth on Issues from Tobacco Smoke to Global Warming* (Bloomsbury, 2010).

28. Knight, "Outrageous Conspiracy Theories."

29. Sunstein and Vermeule, "Conspiracy Theories," 211.

30. José van Dijck, *The Culture of Connectivity: A Critical History of Social Media* (Oxford University Press, 2013).

31. Pierre Levy, *Collective Intelligence* (Basic Books, 1999); Christopher M. Kelty, *Two Bits: The Cultural Significance of Free Software* (Duke University Press, 2008); Clay Shirky, *Here Comes Everybody: The Power of Organizing without Organizations,* reprint (Penguin, 2009); Yochai Benkler, *The Wealth of Networks: How Social Production Transforms Markets and Freedom* (Yale University Press, 2007).

32. Sunstein and Vermeule, "Conspiracy Theories," 204. Owing perhaps to his connections with the Obama administration and perhaps to the specific formulations he puts forth in this article, Sunstein found himself the target of a right-wing conspiracy around the CIA and its role in infiltrating information networks. See Andrew Marantz, "How a Liberal Scholar of Conspiracy Theories Became the Subject of a Right-Wing Conspiracy Theory," *New Yorker,* December 27, 2017, www.newyorker.com/culture/persons-of-interest/how-a-liberal-scholar-of-conspiracy-theories-became-the-subject-of-a-right-wing-conspiracy-theory.

33. Craig Silverman and Lawrence Alexander, "How Teens in the Balkans Are Duping Trump Supporters with Fake News," *BuzzFeed,* November 3, 2016, www.buzzfeed.com/craigsilverman/how-macedonia-became-a-global-hub-for-pro-trump-misinfo?utm_term = .tx3Q91WKl#.sx74NMnBx.

34. Steve Coll, "Donald Trump's 'Fake News' Tactics," *New Yorker,* December 11, 2017, www.newyorker.com/magazine/2017/12/11/donald-trumps-fake-news-tactics.

35. The prime example of existential debate is *Time* magazine's "Is Truth Dead?" cover, which consciously mimicked an equally famous cover from the 1960s, "Is God Dead?" See Pine, "Donald Trump Truth."

36. Lewis H. Lapham, ed., "A History of Fake News," special issue, *Lapham's Quarterly,* 2018.

37. "How 'Fake News' Could Get Even Worse," *The Economist,* July 7, 2017, www.economist.com/the-economist-explains/2017/07/07/how-fake-news-could-get-even-worse; "Lyrebird Is a Voice Mimic for the Fake News Era," *TechCrunch* (blog), April 25, 2017, https://social.techcrunch.com/2017/04/25/lyrebird-is-a-voice-mimic-for-the-fake-news-era/.

38. Charles Acland, *Swift Viewing: The Popular Life of Subliminal Influence* (Duke University Press, 2012).

39. Alongside this popular wave of paranoia around media was the simultaneous emergence of the disciplines of communications research and media studies. This same period is, after all, when seminal scholars like Harold Ennis, Paul Lazarsfeld, and the adherents of the Frankfurt School were turning toward the media as a site investigation for understanding deeper social phenomena. On the whole their response was more measured, even if it benefitted from and in some cases furthered a wider social mistrust. Thus, a scholar like Lazarsfeld, who built his career studying the impact of media, was able to claim, in the early years of communications research, that "mass media are not mainly effective in promoting a specific idea or engendering a stand on a definite issue. What they tend rather to do is to shape for us the picture of the more distant world with which we do not have direct personal contact." Hence, even as newly invented forms of technological media were introduced, political uses (and misuses) were devised, enabling public fear and academic enquiry to go hand in hand. See Elihu Katz, "Communications Research since Lazarsfeld," *Public Opinion Quarterly* 51, no. 4, pt. 2 (Winter 1987): S25–S45, https://doi.org/10.1093/poq/51.4_PART_2.S25.

40. Over the past decade there has been an enormous amount of solid scholarship on these sorts of noncanonical applications of nonfiction film. See, in particular, Karen L. Ishizuka and Patricia Rodden Zimmermann, *Mining the Home Movie: Excavations in Histories and Memories* (University of California Press, 2008); Vinzenz Hediger and Patrick Vonderau, *Films That Work: Industrial Film and the Productivity of Media* (Amsterdam University Press, 2009); and Charles R. Acland and Haidee Wasson, *Useful Cinema* (Duke University Press, 2011).

41. Or, as Bill Nichols might put it, "blurred." Bill Nichols, *Introduction to Documentary* (Indiana University Press, 2001). Propaganda in particular seems to offer a fixed category of criticism for work that differs from one's own rather than a fixed definition. See, for example, the distinctions raised by Jim Leach in relation to Humphrey Jennings. Jim Leach, "The Poetics of Propaganda," in *Documenting the Documentary: Close Readings of Documentary Film and Video,* ed. Barry Keith Grant and Jeannette Sloniowski (Wayne State University Press, 1998), 154–70. Or see the competing interpretations of Grierson's legacy offered in Zoe Druick and Jonathan Kahana, "New Deal Documentary and the North Atlantic Welfare State," in *The Documentary Film Book,* ed. Brian Winston (London: British Film Institute, 2013). See also Dave Saunders, "The Triumph of Observationalism: Direct Cinema in the USA," in *The Documentary Film Book,* ed. Brian Winston (London: British Film Institute, 2013).

42. Roland Marchand, *Advertising the American Dream: Making Way for Modernity, 1920–1940* (University of California Press, 1985).

43. Stuart Ewen, *PR!: A Social History of Spin* (Basic Books, 1996), 34.

44. Stuart Ewen, introduction to *Crystallizing Public Opinion,* by Edward Bernays, reprint (Ig Publishing, 2011), 9.

45. Edward L. Bernays, *Public Relations* (University of Oklahoma Press, 2013); Edward Bernays, *Propaganda,* 1st paperback ed. (Ig Publishing, 2004); Edward Bernays, *Crystallizing Public Opinion,* reprint (Ig Publishing, 2011).

46. Bernays, *Crystallizing Public Opinion,* 25.

47. Ryan Holiday, *Trust Me, I'm Lying: Confessions of a Media Manipulator* (Portfolio, 2013), xiv.

48. Whitney Phillips, *This Is Why We Can't Have Nice Things: Mapping the Relationship Between Online Trolling and Mainstream Culture* (MIT Press, 2015).

49. Even the text's description of its own tactics as a series of "confessions" is an interesting mark of the time in which it emerged. Holiday seems to admit that it is a sin and must be purged from one's conscience if one is to receive absolution, but it is also an invitation to commit the same sin—in the long tradition, perhaps, of Saint Augustine. If further evidence were needed of the connection between Holiday and the broader neoliberal framework of individualized subject that he caters to, one need only look at his output after *Trust Me, I'm lying.* Having reached the age of twenty-three when he published *Trust Me,* Holiday seemed to turn reflective, publishing a series of books that include *The Obstacle Is the Way, Ego Is the Enemy,* and *The Daily Stoic.* These later works fall within the category of individual self-help books focused on the corporate market and executive audience. Holiday's readers, we might assume, are the types of individuals focused not on building someone else's brand or product, but instead on building themselves *as* a brand and a product. Where Bernays was happy enough, or at least self-conscious enough, to distance himself from his reputation as

a "wire puller" and propagandist, Holiday, in the subtitle of his book, avidly embraces this status as part of his self-anointed qualification for the advice he offers. Holiday's title, "media manipulator," is one that might aptly describe both men, but Holiday has the benefit of living in an age where it could be claimed as a point of pride rather than a mark of shame.

50. One of the magazines whose creation Marchand focuses on was *True Story*, which carried the tagline "Is Truth Stranger Than Fiction?" and provided tabloid-style content in a package that could still appeal to the readers of more-respected publications such as *Ladies' Home Journal* and *Good Housekeeping*. See Marchand, *Advertising the American Dream*, 53–55.

51. Michael Schudson, *Discovering the News: A Social History of American Newspapers* (Basic Books, 1981), 93–97.

52. Michele Hilmes, *Radio Voices: American Broadcasting, 1922–1952* (University of Minnesota Press, 1997), 79–82.

53. Daniel J. Boorstin, *The Image: A Guide to Pseudo-Events in America*, 1st Vintage Books ed. (Vintage, 1992).

54. Edward Jay Epstein, *News from Nowhere: Television and the News* (Vintage Books, 1974).

55. Edward S. Herman and Noam Chomsky, *Manufacturing Consent: The Political Economy of the Mass Media* (Pantheon, 2011).

56. Eli Noam's comprehensive data set on media ownership covering a nearly forty-year period also offers a good history of the debate between the libertarians and those like Edwin Baker, a vocal political theorist and legal scholar of freedom and media ownership. See Eli Noam, *Media Ownership and Concentration in America* (Oxford University Press, 2009), 6–15; C. Edwin Baker, *Media, Markets, and Democracy* (Cambridge University Press, 2001); and C. Edwin Baker, *Media Concentration and Democracy: Why Ownership Matters* (Cambridge University Press, 2006).

57. Paul Bond, "Leslie Moonves on Donald Trump: 'It May Not Be Good for America, but It's Damn Good for CBS,'" *Hollywood Reporter*, February 29, 2016, www.hollywoodreporter.com/news/leslie-moonves-donald-trump-may-871464; Alex Weprin, "CBS CEO Les Moonves Clarifies Donald Trump 'Good for CBS' Comment," *POLITICO: On Media* (blog), October 19, 2016, http://politi.co/2el4HVX. Moonves later claimed the comment was simply a misunderstood joke, although many involved in the election (including Marco Rubio, a contender for the Republican nomination at the time) used the comment as proof of the media's efforts to support Trump.

58. Examples of more-focused midstream outlets can be found in categories like politics (*POLITICO)*, food (*Eater*), and style (*The Cut*), not to mention many others.

59. Sarah Ellison, "The Unmanageables," *Vanity Fair: The Hive* (blog), January 2015, www.vanityfair.com/news/2015/01/first-look-media-pierre-omidyar.

60. Wil S. Hylton, "Down the Breitbart Hole," *New York Times*, August 16, 2017, www.nytimes.com/2017/08/16/magazine/breitbart-alt-right-steve-bannon.html.

61. James Risen, "The Biggest Secret: My Life as a New York Times Reporter in the Shadow of the War on Terror," *New York Times: The Intercept* (blog), January 3, 2018, https://theintercept.com/2018/01/03/my-life-as-a-new-york-times-reporter-in-the-shadow-of-the-war-on-terror/.

62. Paul Farhi, "Everybody Wins," *American Journalism Review*, April 2003, https://ajrarchive.org/Article.asp?id=2875; Kevin Coe, David Tewksbury, Bradley J. Bond, Kristin L. Drogos, Robert W. Porter, Ashley Yahn, and Yuanyuan Zhang, "Hostile News: Partisan Use and Perceptions of Cable News Programming," *Journal of Communication* 58, no. 2 (June 2008): 201–19, https://doi.org/10.1111/j.1460-2466.2008.00381.x.

63. Noam, *Media Ownership and Concentration in America*.

64. Yochai Benkler, Robert Faris, Hal Roberts, and Ethan Zuckerman, "Study: Breitbart-Led Right-Wing Media Ecosystem Altered Broader Media Agenda," *Columbia Journalism Review*, March 3, 2017, www.cjr.org/analysis/breitbart-media-trump-harvard-study.php.

65. James Stanyer, "Web 2.0 and the Transformation of News and Journalism," in *Routledge Handbook of Internet Politics*, ed. Andrew Chadwick and Philip N. Howard (Taylor & Francis, 2010).

66. Christopher W. Anderson, "What Aggregators Do: Towards a Networked Concept of Journalistic Expertise in the Digital Age," *Journalism* 14, no. 8 (2013): 1008–23.

67. Scott Shane, "These Are the Ads Russia Bought on Facebook in 2016," *New York Times*, November 15, 2017, www.nytimes.com/2017/11/01/us/politics/russia-2016-election-facebook.html.

68. Paul Arthur, "Jargons of Authenticity: Three American Moments," in *Theorizing Documentary*, ed. Michael Renov (Routledge, 1993), 108–35.

69. Laharee Chatterjee, "New York Times Beats as Digital Subscriptions Surge, Shares Rise," Reuters, February 8, 2018, www.reuters.com/article/us-new-york-times-results/new-york-times-posts-quarterly-loss-as-costs-rise-idUSKBN1FS249.

70. Jean Baudrillard, *Simulacra and Simulation* (University of Michigan Press, 1994).

71. Benkler et al., "Study."

INDEX

Abu Ghraib prison complex: demolition of, 98; as extrajuridical space, 95; *Ghosts of Abu Ghraib* (2007 documentary) (Kennedy), 42; Gregory on, 197n4; leak of images from, 149, 179; photographic images from, 20, 38, 40, 41–45, 48–49; scandal, 38, 50; torture at, 3, 43. *See also Standard Operating Procedure* (2008 documentary) (Morris)

Acland, Charles, 170

activists/protest groups, 50; AIDS activism, 9, 158, 162; *America's Army* and, 88, 99, 110–11, 112; conservative activists, 193n7; cooperation between, 58; Deep Dish TV, 57; documentary as alternative media model for, 12; Fox News and, 64, 66; hackers role in, 182n7; hybridized forms of nonfiction media and, 11, 52–54; independent representations of truth and, 10–12, 183n16; media tradition of, 7; MoveOn-Brave New Films collaboration, 85; new blue media, 57; Newsreel, 57; Obama administration and, 128; Paper Tiger Television, 57; progressive activist left, 157; social media for social change and, 78–81; technology use by, 3–4; transparency and, 179; use of media for social change by, 76; war on terror and, 179; Workers Film and Photo League, 57. *See also* Assange, Julian; Brave New Films; de la Peña, Nonny; *Gone Gitmo* project; MoveOn.org; online digital activism; Weil; WikiLeaks

Adams, Randall Dale, 77

aesthetics: aesthetic change in data processing/display, 118; in *America's Army*, 111; database aesthetics, 13, 44–46, 48, 186n39; data visualization as form of, 127, 137; documentary aesthetics, 4, 6, 8, 9, 10; essential role of, 55–56; in *Gone Gitmo*, 111; image aesthetics in *Standard Operating Procedure*, 37–39; or mash-up videos, 83; Morris use of, 41, 42, 48; in MoveOn videos, 74; in *Outfoxed*, 66; of politics, 10–17, 185n34, 197n34; reenactments and, 42; rhetoric of authenticity and, 178; truth claims and utilization of, 10; universal aesthetic, 124; worldview of audiences and, 182n9

Afghanistan: detention center in, 197n2; invasion of, 59, 97; war in, 2, 20, 21, 55, 127, 144, 145, 149–53; War Logs, 149–53. *See also Rethink Afghanistan* (2008 project) (Greenwald)

Agamben, Giorgio, 94–95

AIDS activism, 9, 158, 162

Ailes, Roger, 64

Alexander, Lawrence, 168

Al Jazeera, 147

Alterman, Eric, 65

AlterNet, 57, 65

Founded in 1893,
UNIVERSITY OF CALIFORNIA PRESS
publishes bold, progressive books and journals
on topics in the arts, humanities, social sciences,
and natural sciences—with a focus on social
justice issues—that inspire thought and action
among readers worldwide.

The UC PRESS FOUNDATION
raises funds to uphold the press's vital role
as an independent, nonprofit publisher, and
receives philanthropic support from a wide
range of individuals and institutions—and from
committed readers like you. To learn more, visit
ucpress.edu/supportus.